NUCLEAR AND TOXIC WASTE

Thomas Streissguth, *Book Editor*

David L. Bender, *Publisher*
Bruno Leone, *Executive Editor*
Bonnie Szumski, *Editorial Director*
Stuart B. Miller, *Managing Editor*

An Opposing Viewpoints® Series

Greenhaven Press, Inc.
San Diego, California

Library of Congress Cataloging-in-Publication Data

Nuclear and toxic waste / Thomas Streissguth, book editor.
        p. cm.—(At issue)
    Includes bibliographical references and index.
    ISBN 0-7377-0475-6 (pbk. : alk. paper)—ISBN 0-7377-0476-4
    (lib. bdg. : alk. paper)
    1. Hazardous wastes—United States. 2. Hazardous waste sites—United States. 3. Radioactive waste disposal—United States.
    I. Streissguth, Thomas, 1958–   II. At issue (San Diego, Calif.)

TD1040 .N83 2001
363.72'87'0973—dc21

                                                          00-030905

© 2001 by Greenhaven Press, Inc., PO Box 289009,
San Diego, CA 92198-9009

Printed in the U.S.A.

Table of Contents

Introduction

In many ways, the 99th Street School, in Niagara Falls, New York, was a very ordinary school. Built in 1955 on land donated by a local business, it held classroom space for about 400 elementary students, who came from the surrounding working-class neighborhood named after an abandoned artificial waterway—Love Canal.

The 99th Street School and Love Canal offered very little of interest to the outside world until 1976, when the State of New York commissioned a private company, the Calspan Corporation, to conduct some tests in the neighborhood. Public officials informed Calspan that the canal had been used as a dumping ground for some 30 years by the Hooker Chemical Company, the firm that had sold the land to the city in 1953. Since that time, Love Canal residents had been complaining of strange events in their yards and basements, and a series of mysterious illnesses affecting their families, and so the public officials had asked for the tests.

Calspan's engineers measured and tested the air, soil, and water around Love Canal, while researchers going through old files and records turned up the fact that more than 20,000 tons of toxic chemicals had been buried in the area. The results of the investigation changed life forever for residents and made Love Canal synonymous with environmental disaster and a pressing modern problem: the disposal of toxic wastes.

The scandal surrounding the contamination of Love Canal was a watershed in the history of environmental action in the United States. An entire neighborhood was evacuated, its homes condemned, its schools closed and its residents moved away. While public officials dragged their feet, alarmed residents held meetings, formed an association, and demanded action. In the next few years, more scandals over toxic waste erupted. Some of them unfolded slowly, over a period of months or years; others happened in a matter of hours, the result of an accident or incompetence. With each newsflash and feature story, a once-unfamiliar word or term—dioxin, plutonium, PVCs, PCBs, DDT, MBTE—entered the ongoing, sometimes bitter public dialogue on environmental issues.

Toxic wastes represent one of the downsides of the industrial progress that has brought convenience, longer life, and more leisure time to the people of the United States. New manufacturing processes use or create as byproducts a wide range of chemicals that are poisonous to humans or animals. These toxins interfere with cell functions, destroy membranes, cause birth defects in newborns, or poison vital organs such as the liver, lungs, or kidneys. Some of them can be metabolized by the body—converted into simpler compounds that can be eliminated. Others do not break down; they accumulate in tissues, bone marrow, vital organs, or the nervous system, causing a range of health problems from mild skin rashes to sudden death.

The laboratory study of toxic chemicals went hand-in-hand with their creation and use in modern industry. In their research, environmental scientists classified these substances according to their effects: neurotoxins, which affect the nervous system; mutagens that cause genetic alterations in cell nuclei; teratogens (which affect the fetus), and carcinogens, the most familiar classification, which includes substances that alter cell reproduction, promoting out-of-control cell growth that can lead to cancer.

Billions of tons of waste are generated each year in the United States, with about 60 million tons classified as "hazardous," meaning it increases the risk of serious—even fatal—illness or pose a substantial threat to the environment. Some of the most dangerous toxic chemicals are chlorinated hydrocarbons, which include polychlorinated biphenyls (PCBs) and DDT, a pesticide, as well as petroleum products that are used in oil refining and in the manufacture of plastics, solvents, and cleaning agents. Also harmful to human health are heavy metals, such as cadmium, lead, mercury, and nickel, which are used in various manufacturing processes.

In the years after World War II, an entirely new industry created an entirely new class of toxic wastes: radioactive elements and isotopes generated by nuclear-weapons manufacturing and the nuclear power industry. Although uranium and plutonium are the best-known components of nuclear waste, there are many others that are dangerous as well, such as the transuranic elements that are created by nuclear reactions. All of these substances pose a danger through their radioactivity, which even in small doses can damage or destroy human tissues and organs. Rather than being simply dumped, radioactive wastes must be stored in a safe place while they degrade, a process that can take hundreds of thousands of years.

More ordinary toxic substances are released into air, water, or soil after dumping by humans, through spills or other accidents, by pesticide spraying, or by careless storage. Water is a particularly good medium for transmission of toxic substances through the environment, since many compounds readily dissolve in water. Rainwater falling on landfills or chemical dumps can cause poisonous runoff that seeps into groundwater or into surface streams, where the toxins gradually spread out into the environment and are absorbed by plants through their roots, stems, or leaves. Animals eat the plants or drink the water contaminated by toxic wastes; humans ingest them by breathing the air, drinking the water, or eating contaminated animals or plants.

Rules and regulations

Love Canal and other toxic crises intensified what had been a mild debate over the best way to deal with toxic and nuclear wastes. As with other such issues, the debate eventually attracted the attention of elected officials, who reacted by passing legislation intended to control or eliminate the problem. Under the National Environmental Policy Act, passed by Congress in 1969, standardized rules were established governing the identification and testing of toxic substances. In 1976, the United States Congress passed the Toxic Substances Control Act (TSCA), in many ways the forerunner of modern toxic waste law. This law authorized a federal agency, the Environmental

Protection Agency (EPA), to monitor and control artificial chemicals. The TSCA requires that the EPA be notified before any new chemical is manufactured, and it allows the EPA to regulate the production, use, testing, and disposal of chemicals. The EPA also has the authority to control 65,000 existing chemicals that were already in use before the act was passed.

More legislation followed. The 1976 Resource Conservation and Recovery Act set up standard policy for regulating toxic and hazardous substances, from production to disposal. The 1977 Clean Water Act focused on toxic wastes in water. The 1980 Comprehensive Environmental Response, Compensation, and Liability Act (CERCLA) set up the Superfund program to deal with abandoned toxic waste sites. The 1990 Clean Air Act seeks to reduce toxic emissions into the atmosphere.

Individual states have also passed laws intended to reduce toxic waste. These laws set specific goals for polluters to reduce their production of toxic chemicals by a certain percentage over a certain period of time. Many also require companies to inventory their toxic chemicals and keep careful records of their use and disposal.

Controversies

Such laws can quickly go out of date. New chemicals are being invented and put into use every day in the United States, with about 1,000 new chemicals introduced each year. There are 100,000 substances officially listed as toxic and/or hazardous by the National Institute of Safety and Health, and one-third of the 700,000 different chemicals in current use have never been tested for their effects on human health.

Most of the laws designed to regulate these substances have the support of advocates who believe that strict regulation of chemicals and their manufacturers will bring about a lessening of toxic waste dangers. Environmental activists say that business owners are motivated by profits, not by better public health, and that the only way to guarantee their cooperation is to pass new laws and regulations. Many activists want to see stricter controls over certain toxic chemicals or outright banning of some substances, as was done with pesticides such as DDT. They want to see public or private funding of new technologies that rely on "clean" energies (solar, wind, or thermal) or that carry out the complete recycling of toxic waste. They also advocate better communication and cooperation between private firms and the communities where those firms do business.

On the other hand, environmental laws are never written and passed without opposition, especially from manufacturers, who claim that such laws are bad for business and make it harder for them to provide important economic benefits to the communities where they are located. They point out that industries that are unhindered by regulation and central control provide the cleanest technology and that free-market solutions are the most effective means of reducing environmental contamination.

Future wastes

In the early years of the twenty-first century, one focus of the debate over toxic wastes will likely be the nuclear power industry. In the next few years, Congress must make a decision on the treatment of nuclear wastes:

whether to wait for a practical solution for "on-site" storage, or to ship the nation's entire inventory of radioactive waste to a central dumping ground located at Yucca Mountain, in the high desert of Nevada. The debate will focus public attention on the nuclear waste issue in particular and on toxic waste in general.

There is one certainty in the debate over toxic and nuclear wastes: the debate will go on for a long time, and fiercely. Hundreds of organizations have been formed by advocates of one viewpoint or the other in order to present their case to the public and the government. Books and magazines devoted to the subject abound. Television programs and videos are produced, radio programs are aired, and World Wide Web pages are created in order to inform and persuade. The issue is ongoing and divisive because toxic waste can threaten individuals, families, and entire communities, giving an unpleasant counterpoint to the time-honored national faith in the benefits of invention, industry, and scientific progress.

1

The Superfund Program Should Be Eliminated

James DeLong

James DeLong is an attorney specializing in environmental issues.

The federal Superfund program has been embroiled in controversy ever since it was established as part of the Comprehensive Environmental Response, Compensation, and Liability Act (CERCLA) of 1980. Although it has achieved some successes, the program has been largely ineffective in its goal of cleaning up the nation's worst toxic waste dumps. As James DeLong explains in a 1997 article written for *Regulation*, a journal published by a Washingon, D.C.–based think tank known as the Cato Institute, Superfund mandates joint responsibility among all those even remotely identified as contributing to each toxic site. The result has been a vast tangle of lawsuits and litigation that has put large fees in the pockets of lawyers but paralyzed the cleanup and remediation of poisoned land and water. Although a reform of the Superfund law was proposed in January 1997 by Senator Bob Smith (the Superfund Cleanup Acceleration Act), the law has yet to be passed.

DeLong argues that the solution to Superfund's problems is not more legislation but rather to turn over the program's functions to state and local authorities and abandon the Superfund program altogether.

The Superfund program was established in 1980 to clean up hazardous waste dumps. Since its inception, the program has spent some $40 billion from general federal revenues as well as from a special tax on the chemical and petroleum industries. Senator Bob Smith (R-N.H.), Chairman of the Superfund Subcommittee of the Committee on the Environment and Public Works, recently endorsed the comment of an unnamed former EPA administrator who said that Superfund is "the worst program ever enacted by the U.S. Congress." He announced an intent to "restructure the program from top to bottom." To that end, he introduced

Reprinted from "Inadequate Superfund Reform," by James DeLong, *Regulation*, Spring 1997. Reprinted with permission from The Cato Institute.

the Superfund Cleanup Acceleration Act of 1997 (S. 8). Unfortunately, while the bill will provide some relief from some of the law's worst provisions, Superfund is fundamentally defective. Even a far more radical "restructuring" than S. 8 would be of only minor use. The program needs to be uprooted.

Fundamental flaws

For the entire seventeen years of its operation, the Superfund program's basic flaws have been evident. First, the program imposes liability for cleaning up hazardous substances on everyone connected with the pollution, including the generators of the alleged pollutants, the transporters, and the landowners. It even extends liability to parties who lent money and became active in the affairs of the borrower. Further, Superfund makes liability retroactive for events that occurred before the law establishing the program was passed.

A second major problem with Superfund is that liability is strict. It does not distinguish according to amount of waste. The rule of joint and several liability applies. A party that discards one pound of waste in a dump the size of Manhattan could be billed for the cost of cleaning up the entire dump. The liability extends to individuals who have their waste picked up by municipal trash haulers, who in turn dispose it as directed by local public officials.

Third, cleanup standards are absurdly stringent. Congress's preference for "permanent" solutions has been interpreted as requiring cleanup rather than containment, regardless of cost or benefit. Future use of the land and the economic value of the property are not taken into account in the decision. Thus it might cost $30 million (the average price of cleaning up a major site) to clean up a piece of property in an old industrial park to residential standards even though the property will simply be used as another industrial park in the future. To compound the problem, "worst first" decrees require restoration of the most polluted sites before restoration of sites that might be made usable again with minimal expenditures.

The scope of Superfund's coverage is unmanageable. EPA has designated about 1,250 sites on its National Priorities List (NPL). . . . Yet another thirty thousand sites, rejected for inclusion in the NPL, are still covered by the basic Superfund liability laws. They can be the subject of private lawsuits.

High transaction costs is a fourth costly effect of Superfund. At least 30 percent of Superfund expenditures from government and businesses are spent on the legal process, not on clean-up. Because of the high stakes, companies have strong incentives to challenge Superfund efforts. The fees of lawyers and consultants mount. For most sites, cleanup costs are divided according to complicated multi-factor equitable tests that invite wrangling.

Since everyone who ever used the site is liable, the number of parties brought into the game expands. The EPA sued eleven companies who in turn brought in 180 more, and those added another 590. Swept up in the last tier was the owner of a diner who swore her contribution consisted solely of mashed potatoes and similar restaurant waste. Her story is distressingly typical.

The EPA blames the corporations for such abuses. That defense misses a fundamental point. It is the statute, written by Congress with help from the EPA, that makes the restaurant-owner liable. The Agency's desire to keep everyone legally responsible and then exercise complete discretion to decide who actually pays makes the whole situation worse, not better.

Disposal of hazardous waste is simply not a federal problem. Sites are in particular states and should be the responsibility of those states.

Lastly, the scope of Superfund's coverage is unmanageable. EPA has designated about 1,250 sites on its National Priorities List (NPL). Those are the only sites in which EPA is directly involved. Yet another thirty thousand sites, rejected for inclusion in the NPL, are still covered by the basic Superfund liability laws. They can be the subject of private lawsuits. States have created their own programs that address tens of thousands more sites. In fact, every site where any amount of chemicals has been leaked, spilled, or dumped is covered by Superfund. The total must run into the hundreds of thousands, maybe the millions. Most could be cleaned up with minimal effort. But the program pressures new industries to head for the suburbs, for the "greenfields," rather than to redevelop old urban industrial plots, usually referred to as "brownfields."

Reform efforts

S. 8 attempts to alleviate Superfund's most obvious defects. First, it would exempt some future innocent purchasers of property from Superfund liability. Further, for NPL sites, it would eliminate the liability of minor parties, small businesses, and municipalities.

Second, S. 8 reforms the cleanup standards, potentially allowing containment rather than complete clean-up at some sites. Specifically, EPA risk assessments would be required to consider the future uses of the land when deciding between clean-up or containment. Skeptics might question the extent of sound judgment EPA will exercise in making those evaluations. But at least the reform would prevent EPA from claiming that the statutes preclude more economically responsible options.

Third, S. 8 replaces the judicial process of allocating costs among the parties with an administrative one. But the standards remain amorphous, and it is not clear whether bureaucrats would do a fairer job than judges in making such decisions.

S. 8 further makes a bow towards federalism, giving the states some decision-making authority over NPL sites. But that devolution is really a

Potemkin Village. States must ask for EPA's approval of their decisions every step of the way.

S. 8 deals with the brownflelds problem in a manner all too common for government. Rather than eliminating economically destructive regulations, the bill creates a $40 million slush fund to be doled out to states and cities for cleanup. Further, the funds will be doled out only after a detailed process of planning and review, wasting significant amounts of money along the way.

Finally, S. 8 offers special breaks, such as liabilities exemptions, for certain interest groups. The groups include clean-up contractors; recyclers; railroads; and educational, scientific, charitable, and religious organizations.

Not good enough

But the same question EPA asks concerning Superfund sites (Should they be contained or eliminated completely?) should be asked of the Superfund program itself. S. 8 merely tries to contain a bad program. But its toxic effects on the economy necessitate its elimination.

First, disposal of hazardous waste is simply not a federal problem. Sites are in particular states and should be the responsibility of those states. A federal response capability for certain kinds of emergencies might be defended on the grounds of the economies of scale involved. Or, if waste is dumped in a river, then some form of interstate and thus, perhaps, federal action might be needed. Even in such cases caution is necessary. Interstate groundwater contamination is often used as the excuse for federal intrusion, but the documentation of serious problems in that area is exceedingly thin. Most waste goes into the ground, where it just sits. Its effects are not interstate and offer no grounds for federal involvement.

A second major shortcoming of S. 8 is that it applies principally to NPL sites in which the EPA is directly involved, not to other sites that are still covered by Superfund regulations. Private suits are still permissible under the program's complex rules and other costly actions could still be required.

A third problem with the legislation is that it leaves a huge loophole that could negate reform attempts. S. 8 requires EPA risk assessment studies to consider future uses of land. Agreed, future use of land must be taken into account, but the cost of cleanup should drive the land-use decisions, not vice versa. Herein lies the problem: the EPA can predict any pollyannaish uses for the land they choose. A site to be used as waste dump needs not meet the same standards that might be appropriate for a kindergarten.

Though well-meaning, S. 8 merely takes a few shovelfulls of sludge from a toxic program that cannot be reclaimed. The debate should be not over how to reform Superfund but, rather, over how to eliminate it.

2

The Superfund Program Can Operate Successfully

The Environmental Protection Agency

The Superfund program faced one of its biggest challenges at the site of the Anaconda copper-smelting operation in Montana's Deer Lodge County. Beginning in 1884, workers at the Anaconda mine dug and smelted an immense amount of copper ore, providing an economic boom to the region but leaving more than 1,500 acres in the surrounding hills and valleys denuded of vegetation and contaminated with toxic waste.

The Environmental Protection Agency, which operates the Superfund program, points to the Old Works Golf Course at Anaconda as one of its success stories. In partnership with ARCO, the property owner and the corporation responsible for Anaconda's mining operations, a plan was undertaken in 1994 to clean up a 250-acre portion of the site and transform it into a golf course (to be designed by golf champion Jack Nicklaus) and tourist destination. The Old Works Golf Course demonstrates that, when properly planned and carried out, Superfund programs can revitalize industrial and mining regions now threatened with decline in the face of a changing "high-tech" economy.

A slice on a hole at the Old Works Golf Course in Anaconda, Montana, could leave golfers with a tough shot out of a nearby slag trap. That's right, slag. When golf great Jack Nicklaus helped design the course on this former copper smelting and processing area, he chose to make the bunkers out of the charred mining wastes that remained on the property. He also opted to incorporate old mining artifacts in the course's overall layout, so that golfers play in sight of old smelting ladles and chip in sight of flues and smelting ovens. This unusual course is part of the dramatic transformation of the Old Works/Anaconda Smelter Superfund site. It also is the centerpiece of Anaconda's transformation from a mining town to a recreational and tourist destination. What follows is the story of how EPA worked with others to clean up and return the site to productive use,

Reprinted from the Environmental Protection Agency Superfund website, www.epa.gov/superfund/programs/recycle/casestud/anacocsi.htm.

13

and of the positive economic impacts and environmental and social benefits that resulted.

A site snapshot

Anaconda Smelter was once one of the shining stars of the American mining industry. It employed thousands of people and, as in most mining towns, was the backbone of Anaconda's local economy. Located in the foothills of Montana's Anaconda-Pintler Mountain Range in Deer Lodge County, the facility first began copper smelting operations in 1884. The smelter rose quickly to national prominence because of its astounding annual copper production and exhaustive exploitation of copper from the surrounding area. But this all came at a price to the environment. The land was left gouged with mines and extensively contaminated with heavy metals. Today, about 10,500 people live within two miles of the site. To the north and east, the site is bound by Warm Springs Creek, to the south by the Anaconda-Pintler Mountain Range, and to the northwest by the Flint Creek Mountains.

Anaconda Smelter was once one of the shining stars of the American mining industry. It employed thousands of people and, as in most mining towns, was the backbone of Anaconda's local economy.

From smelting ladles . . .

Because of Anaconda Smelter's size and the large amount of waste remaining on the property, EPA divided the site into 11 smaller units. One of the units, the 1,500-acre Old Works/East Anaconda Development Area, was used as the primary smelting and processing area from 1880 to 1902. The area contained over a million and a half cubic yards of soil, slag, and flue dust and was contaminated with a range of pollutants. EPA worked closely with ARCO (the owner of the property) and the community to develop a cleanup plan that not only protected people and the environment, but also preserved its historical significance and allowed for redevelopment. What this partnership unveiled in 1994 was a plan that included cleaning up the contamination and building a top-notch golf course over a portion of the area. Under EPA supervision, more than 200 ARCO employees covered the 250-acre area designated for the course. They covered the contaminants with 18 to 20 inches of soil, revegetated the area, and installed a state-of-the-art drainage system. Other portions of the site were also covered with clean soil, and the embankments along Warm Springs Creek were upgraded.

. . . To golf clubs

To design the course, ARCO enlisted the help of golfing legend Jack Nicklaus, who had already designed golf courses on waste dumps in Michigan and lava fields in Hawaii. Nicklaus took advantage of the area's beautiful mountain vistas and preserved many of the unique historic characteristics of the former smelting site. For example, the bunkers on the course are filled with more than 14,000 cubic yards of inert smelting slag ground to the texture of sand. Even the course's drainage system recycles water. Excess rainwater from the elaborate system is channeled to a holding pond and reused to irrigate the grounds, making it an environmentally-friendly golf course. To top it off, a historic hiking trail now winds its way around the course. The trail highlights Anaconda's smelting heritage and gives hikers an insight into copper mining techniques of years past.

Community benefits

The town of Anaconda is redefining itself. Originally a mining town, it has turned to tourism and recreational pursuits to boost its coffers and provide jobs for its citizens. The golf course at Anaconda Smelter is a critical step in this direction. The cleanup and redevelopment efforts have already brought about dramatic economic impacts and social and environmental benefits. The temporary and full-time jobs created to complete the cleanup and construct and operate the golf course have injected the local economy with new revenue and provided the state with additional income and sales taxes. The golf course has proven to be a tourist magnet, attracting out-of-towners to the area who enjoy golfing, skiing, fishing, and hunting. This has had an immediate and positive impact on property values at and around the site. Numerous developers have expressed interest in purchasing land near the course, and more than 50 existing businesses have made improvements to their properties. Ultimately, renewed interest in the area will serve to help Anaconda transform itself into a popular recreational vacation spot. The barren and battered landscape is slowly being restored to its former beauty. Trout once again fill Warm Springs Creek, and the plant and animal life are flourishing.

The bunkers on the course are filled with more than 14,000 cubic yards of inert smelting slag ground to the texture of sand.

Keys to success

The strong partnership forged among EPA, ARCO, and the local community was the key ingredient to the successful turnaround at the Anaconda Smelter site. Both ARCO and the local community played an active role in helping EPA plan the cleanup and redevelopment. EPA also helped orchestrate an agreement between ARCO and Deer Lodge County that transferred ownership of the golf course to the county. The agreement required ARCO to maintain the integrity of the cap and ensure that deeds for the property include restrictions that prevent disrupting the cap. In addition, Deer Lodge County agreed that all revenues from the golf course not used to cover operating expenses would support the historical preservation of the property and hiking trail and go toward continued economic and recreational development within the county. This spirit of cooperation has helped put Anaconda back on the road to recovery.

3

Industrial Pollution Increases Cancer Rates

Barbara Koeppel

Barbara Koeppel is a Washington, D.C.–based investigative reporter.

The region known as "Cancer Alley," which lies along the Mississippi River valley in eastern Louisiana, has some of the highest rates of cancer, miscarriages, neurological diseases, and other health problems in the nation, yet the industrial plants and waste-processing companies present in the area deny that their operations are the cause. In the meantime, these businesses work hand-in-glove with the state government to assure that expensive, preventive, and remedial measures do not become law.

Something is rotten in the state of Louisiana. It is the stretch along the Mississippi River between Baton Rouge and just south of New Orleans. Locals call it Cancer Alley. The corridor is home to seven oil refineries and somewhere between 175 and 350 heavy industrial plants, depending on how you count. Together, they produce staggering amounts of waste, much of which they treat on-site or spew into the air, land and water. Waste-processing companies also set up shop here to handle the industrial overflow. Because the laws are laxer in Louisiana than elsewhere, they cart in even more waste from outside the state to bolster business—a whopping 330 million pounds in 1993, which dropped to about 52 million pounds in 1995, partly because the federal government closed down Marine Shale, one of the state's three waste-processing companies.

People living nearest the factories and waste dumps are sick and dying. Clusters of asthma, stillbirths, miscarriages, neurological diseases and cancers have mushroomed. And residents have long claimed that the waste has poisoned domestic animals, wildlife and fish. Children, because they're still growing rapidly, are at greatest risk. Eight-year-old Caleb Thomas and his family know this well. Until he was 6, they lived in Gonzales (pop. 18,000), which sits in the shadow of five petrochemical plants and several waste dumps. One day just after Caleb's second birthday, he ran into a wall inside his house. He fell again and again, and threw up until there was nothing

Reprinted from "Cancer Alley," by Barbara Koeppel, *The Nation*, November 8, 1999. Reprinted with permission from the November 8, 1999, issue of *The Nation*.

left. Soon, the right side of his face was paralyzed, his eyes were crossed and an ear infection raged. Within a couple of weeks, the sentence was pronounced: Caleb's small body was being ravaged by rhabdomyosarcoma, a rare childhood cancer. For two months, he was shot full of chemicals and irradiated daily, which burned his head, face, inner ear and mouth. The sores in his mouth were so bad he couldn't eat for a year and had to be fed through tubes.

Today, Caleb's therapy seems to have worked. The baseball sized tumor in his brain disappeared, as did the cancer cells in his spine. He plays football, goes fishing and wants to be a baseball player, doctor or police man. But the cure, if indeed it is one, has come dear. Radiation ruined Caleb's pituitary gland, so his father, Sam, must inject him daily with hormones or he won't grow. He is deaf in his right ear, which affects his speech. And radiation collapsed his jawbones, so he can't open his mouth wide enough even to eat a piece of fruit.

"Children are very susceptible to pollution, so they're the first to get sick and are an early warning of what can happen wholesale if the course is not reversed."

Caleb's type of cancer should hit just one in a million US children a year. But in Gonzales, cancer is no rarity. Two other boys were also diagnosed with rhabdomyosarcoma, bringing to three the total who fell ill within fourteen months. In Zachary, forty miles away, four more had developed it a few years earlier. All have died. "Everybody knows someone whose child has had cancer," says Michele Thomas, Caleb's mother. Her cousin died of leukemia when he was 10. Today, five more children in or near Gonzales have leukemia.

Deadly cancers

The scene is the same all along the corridor and in parishes (counties) nearby. In Denham Springs and Walker (combined pop. 14,000), six children who lived within three miles of an oil and chemical waste dump that operated in the sixties and seventies developed neuroblastomas, deadly cancers of the central nervous system—a rate 120–200 times higher than the norm. Six children were diagnosed with leukemia from 1981 to 1990, while one developed a medulloblastoma (brain tumor) and another a chondrosarcoma (bone-cartilage cancer). In one household, besides the child with leukemia, four family members also died of cancer.

In Morgan City (pop. 14,000), pediatrician Anthony Saleme says five children were diagnosed with neuroblastomas in 1986–87 and a sixth in the early nineties (all have died), and one brain tumor and three leukemias were diagnosed from 1989 to 1993—more than he had seen in all the years since his practice started there in 1974. All the cases developed after Marine Shale opened in 1984.

And the cancers don't hit only humans. Biologist Florence Robinson, who lives just outside Baton Rouge, put three dogs to sleep in 1992, two with breast cancer and one with liver cancer. In 1996 her fourth dog died of breast cancer. No other dogs from the same litters developed cancer. Nor did they live in Louisiana.

Some doctors share the residents' concerns. Dr. Floyd Roberts, Baton Rouge lung specialist, says many of his asthma patients never had the disease until they moved to the area. Moreover, most say their symptoms disappear when they leave Louisiana and reappear when they return. Dr. Judd Patten, an oncologist who moved to Baton Rouge in the early nineties, says he saw far more cancer than he expected. He also found that the preferred chemotherapy for treating lymphoma, which generally succeeds about 65 percent of the time on a national basis, didn't work with his Louisiana patients. Patten wonders if these were chemically induced lymphomas and thus resistant to chemical treatment.

Dr. Patricia Williams, director of the Occupational Toxicology Outreach Program at Louisiana State University Medical Center, says health officials downplay the importance of the childhood cancer clusters. She believes they function like canaries in a mine. "Children are very susceptible to pollution, so they're the first to get sick and are an early warning of what can happen wholesale if the course is not reversed," she cautions.

Debating the tumor registry

Until now, those who believe Louisiana's toxic gumbo creates the cancers that kill their children haven't been able to prove it. Their nemesis is the state's Tumor Registry, which is supposed to record all cancer cases and which officials tout as something akin to the Bible. According to its figures, only Louisiana's white male lung cancer rates are significantly higher than those in other states; white females' lung cancer rates are a bit higher.

Dr. Vivien Chen, the registry director, admits that Louisianans die from cancer at a higher rate than do people in other states, but she attributes this to poverty, which prevents many residents from getting timely medical care. She also blames lifestyle—insisting that residents along the corridor smoke earlier and more over their lifetimes and eat a Cajun diet heavy on fats and short on fruits and vegetables. However, in 1997 the federal Centers for Disease Control in Atlanta placed Louisiana in a tie with two other states for fifteenth place, among all fifty, in the number of residents who smoke. And since lung cancer is most often fatal, the issue of early and reliable treatment is almost irrelevant. As to the lifestyle line, residents don't buy it, since much of the state's population is either rural or in small to medium-sized towns, and most homes have vegetable gardens. Regarding the cancer clusters, Chen says: "Purely random. If you toss five coins and they land heads up, this doesn't mean they only have sides with heads. Similarly, five neuroblastomas in a small town doesn't mean they are linked to pollution."

Soon after the movie *A Civil Action* was released, articles discrediting the validity of the link between pollution and cancer clusters ran in the *New York Times*, *The New Yorker* and the *Wall Street Journal*. "The problem is that reporters mostly rely on industry-supported scientists, who naturally take this line. Clearly, these articles are related to the movie's release, which raises the cluster specter," says Bill Ravanesi, the Boston project director of Health Care Without Harm, a coalition of 185 health professionals, unions, environmentalists and others.

Critics, however, claim the clusters are hardly chance occurrences. Six former state Tumor Registry employees and current registrars at local hospitals insist that the way the numbers are collected and combined dramatically conceals the facts. None agreed to be identified, and a seventh refused to talk, even off the record. "I have to live and work here, so the subject is closed," she said, hanging up the phone. The biggest problem is that the registry divides numbers from various parishes, some industrial, some rural and some mostly uninhabited swamp, among ten regions. What this means, says Williams, is that it's impossible to see what happens close to the most polluted sites. "This is what we need to do to make the link between pollution and the diseases," she says. Moreover, the six registrars question the registry's accuracy, citing sloppy recording techniques.

Williams contends that the clusters develop as they do because people's exposure to chemicals varies greatly. This level of exposure depends not only on the distance they live from the waste dumps or petrochemical plants but also on where they work and where their children play, the direction of the wind, the types of chemicals emitted, the length of time the wastes accumulate, the flow of underground or surface water, whether local creeks flood into vegetable gardens, how often residents eat local fish and game, and whether drinking water comes from wells or the Mississippi.

Some of the worst polluters donate lavishly to all sorts of institutions and are particularly generous to environmental causes.

Sam Thomas knows this firsthand: When his children and cousins swam in a creek near their home, he says, they all got earaches. When they swam in a different one, nothing happened.

Chen says it would be too costly to report the numbers by smaller geographical areas. However, one current and one former regional registry director say that until recently their units, which funnel numbers to the central unit, collected data by census tracts (which are smaller than postal zones), but then the state lumped the numbers into larger, regional units. Williams says the registry also makes it tough to expose clusters, because instead of listing cancers by the type of cell involved—which scientists need to know to document the cancer's incidence—it records only the body sites where tumors appear, such as the lung or spine.

Feeding heavy

Another issue is the way researchers study the clusters. Williams stresses that unless they look at the small picture, they will never understand the link between disease clusters and chemical exposure. Until now, health studies in Louisiana have done exactly the opposite. Thus, as in a 1988 LSU [Louisiana State University] and state health department study of Morgan City's neuroblastomas, researchers always find "nothing conclusive."

Wilma Subra, a Louisiana chemist and adviser to the National Commission on Superfunds who recently won a MacArthur fellowship, says the Morgan City study never investigated how Marine Shale, the waste-processing company, disposed of its chemicals from the time the children with neuroblastomas were conceived until they were diagnosed. Nor did researchers interview former employees, who admitted in a federal case against the company (for violating emissions regulations) that Marine Shale officials instructed them to remove the incinerator scrubbers at night and "feed heavy." Subra says this reveals that Marine Shale, to save on costs, burned its waste without filters and discharged far more than it recorded. She adds that it got away with this because equipment that monitors emissions operates only in daylight.

Other studies were equally faulty. In 1993, when the state investigated cancer clusters near the Denham Springs/Walker dump site, it concluded that exposure to the chemicals at the site was too low to cause adverse health effects. The study identified only eight leukemia cases, including just one in a child (which it said was not unusual), plus one neuroblastoma and five other cancers in adults. Williams, who studied the case for 9,000 plaintiffs suing the companies that used the dump, found six children between 1981 and 1990 and eleven adults between 1968 and 1994 diagnosed with leukemia. From 1970 to 1993 six neuroblastomas, as noted, a medulloblastoma and one chondrosarcoma were diagnosed in children under 18. While the state relied on the registry numbers, Williams used the plaintiffs' documents and medical records. Also, she interviewed families and found striking similarities in what the children ate and where they played or went to school. She concluded that the much larger number of cases, plus the sources of exposure, were enough to connect the site to the cancers. In 1997 the companies settled for $131 million.

Industry's donations

Perhaps the biggest obstacle to proving the pollution-cancer link is the industrys' immense wealth and influence. In 1998 oil, petrochemical products and gas extraction accounted for $28 billion of the state's $110 billion gross state product. Some of the worst polluters donate lavishly to all sorts of institutions and are particularly generous to environmental causes. According to press reports, Freeport McMoRan, one of the world's largest manufacturers of chemical fertilizers and the major discharger of toxic waste into Louisiana waters in the early nineties, donated $5 million to the Audubon Institute to build a wilderness park for endangered animals, $2.5 million to LSU to create the Institute for Recyclable Materials and another $1 million for the university's cancer center, $1.6 million to the University of New Orleans Center for Environmental Modeling and $350,000 to the Nature Conservancy to preserve wetlands, to name a few, over a time period during which the administration of Governor Buddy Roemer was prosecuting it for illegal dumping. In 1998 Shell gave $5 million to the National Fish and Wildlife Foundation "for conservation of the Gulf ecosystem." Exxon gave $30,000 to protect 425 acres of hills, while Texaco sponsored volunteer tree planting.

Most troubling, however, are the massive gifts these industries make to the universities and medical centers, which, together with the state health department, run the cancer studies. According to press accounts, oilman C.B. Pennington gave LSU $125 million in the eighties to build a Bio-Medical Research Center, whose main task is to study nutrition; and Lod Cook, chairman of ARCO, footed most of the bill for LSU's three-story alumni center, which opened in the mid-nineties. In August of 1997, when Pennington died, he left about $250 million to be shared among the LSU center, the Pennington Foundation and his grandchildren.

Tulane has also thrived, according to Earl Bihlmeyer, senior associate vice president of the university. Texaco donated a twenty-year free lease for a building that houses its Public Health School facility. And because the lease will soon expire, Tidewater Industries, which services oil rigs, has given it a twenty-four-story building that will also be used by Tulane's medical and hospital departments. In 1996 Freeport donated $1 million to the university's Bio-Environmental Research Center and, along with Shell and Exxon, pumped another $2 million into its Environmental and Waste Management Program. In fact, dollars flow for countless projects, such as endowed chairs, which cost donors $600,000 each, matched by $400,000 from the Oil and Gas Trust Fund (financed by gas and oil royalties). Since the early nineties, Ethyl, Texaco and Claiborne Gasoline have each endowed an LSU chair, Freeport has endowed one Tulane chair and two LSU chairs, and Pennington has given two chairs to Tulane.

Political investments

The problem, say critics, is that contributions profoundly affect the kinds of studies designed: Until the mid-nineties, for example, scientists investigating lung cancer limited research to the link with smoking, says Ben Fontaine, executive director of Louisiana's American Lung Association. Efforts by journalists and others to get the universities to reveal their funding sources (apart from data about endowed chairs) have been stonewalled: Tulane's status as a private institution allows it to remain silent, and although LSU is a public university, it created a private foundation through which it funnels its grants.

The industries and firms servicing them also contribute heavily to politicians on both sides of the aisle. For example, oil and gas companies contributed just under $342,000 to the 1990 campaign of Democratic Senator J. Bennett Johnston, chairman of the Energy and Natural Resources Committee. In the early nineties, the committee again exempted oilfield waste—which contains carcinogens, heavy metals and radioactive materials—from federal hazardous-waste laws (the first time it was classified as nonhazardous was in 1970). Thus, oil companies can inject it underground or just dump it into pits, with far fewer controls. Similarly, Louisiana state legislators got $294,000 in 1993–94 from the industries. Soon after, the lawmakers killed a bill that would have sharply hiked industry taxes.

Sometimes, the links between industry and government are even closer: Republican Mike Foster, the current governor, who owns prime oil and gas land, earns $200,000 in annual royalties from Exxon, along with smaller sums from Quintaine Petroleum, Meridian Oil and others, according to his campaign disclosure form. Moreover, according to press reports, firms such as Chevron, Occidental, Ciba-Geigy, Freeport and Cryptopolymers were among a long list of industry contributors to Foster's 1996 gubernatorial campaign. Edwin Edwards, who held the governor's chair in 1972–80, 1984–88 and 1992–96, earned, in the years he was out of office, at least $100,000 a year in oil and gas royalties from Exxon, Superior Oil and others, and large legal retainers from several oil corporations, including Texaco and Texas International.

Instead of banning fishing in contaminated waters, [Louisiana] erects a few warning signs. It also publishes a brochure instructing pregnant women on how to cook contaminated fish . . .

Some legislators are even employed by the industry at the same time they hold office, while others belong to law firms that represent the corporations. Also, top civil servants shift easily from the public to the private domain. For example, Jim Porter, who from 1984 to 1988 was the director of the state's Department of Natural Resources, which promotes the oil industry, took the top slot at its trade association, the Mid-Continent Oil and Gas Association, a year and a half after leaving public office.

Such investments appear to have paid off. When geologists at a state agency testified in the early nineties that lands owned by officials' friends should not be used for underground-injection wells or landfills because they were too close to groundwater sources, they were told never to testify again. Soon after Edwards regained the governorship in 1992, he cut off the unit's funding. After Tulane's Environmental Law Clinic agreed in 1996 to represent the mostly poor and black residents in the already highly polluted St. James Parish who were tying to block Shintec from building a large polyvinyl chloride plastics factory, Governor Foster blasted the clinic on television, threatening to reduce its tax breaks.

Living with toxic waste

Indeed, the state does all it can to hide the extent of pollution. Instead of banning fishing in contaminated waters, it erects a few warning signs. It also publishes a brochure instructing pregnant women on how to cook contaminated fish and to eat no more than two servings of certain species a month. Such avoidance techniques should worry non-Louisianans, too, since about 20 percent of oysters and 7.5 percent of shrimp consumed in the United States come from Louisiana waters. (The shrimp figures were far higher before foreign imports increased over the past decade.)

Toxic emissions declined dramatically from 1988 through the early nineties, as a result of the tightening of regulations during Governor Roemer's administration, but they rose soon after industry helped defeat Roemer in the 1992 election. By 1995, according to the federal Environmental Protection Agency, the state was number one among all fifty states in dumping into waterways and second only to Texas in its combined discharges into water, air and land, and into underground-injection wells. But because Louisiana is only a sixth the size of Texas, the pollution here is far more concentrated.

An environmental scorecard

During the Roemer years, Dr. Paul Templet, an environmental scientist and then head of the state's Department of Environmental Quality [DEQ], got the companies to clean up their acts. For example, he introduced an environmental scorecard: Firms that did not reduce emissions lost their property-tax exemptions, which had been routinely approved. "It's naïve to think the industries will control pollution voluntarily," Templet says. Firms also had to submit environmental assessments when they wanted to expand, those injecting oil waste into wells had to find other ways to dispose of it, and waste processors were heavily taxed on what they imported into the state. When companies refused to comply, the DEQ got serious about penalties: These jumped from $1.5 million in 1988 to $7.6 million in 1989.

Templet says that while fines this size are minuscule to such corporate giants, they harm their reputations, which counts when they want to locate facilities in other areas. As a result, during his tenure firms upped their pollution-control spending by 600 percent, recycled more of their waste and cut their imports of toxic waste by nearly half. Emissions dropped by 50 percent.

But the gains were brief. Once Edwards returned in 1992, his administration scrapped the environmental scorecard, slashed the tax on imported waste and whittled penalties down to under $2 million—all within the first year. Beginning in 1996, Governor Foster finished the job where Edwards had left off. Among other actions, he limited the requirement for environmental assessments, which now need to be made only for "major," facilities—and the DEQ decides what's major. Penalties dropped even lower: Although Tim Knight, administrator of environmental technology, insists that the DEQ levies up to $50,000 a day on firms that have not complied with the laws within the time allotted, Bob Kuehn, former director of the Tulane Environmental Law Clinic, says DEQ penalties in 1997 totaled just $736,000, the lowest since 1988, with fines averaging about $17,500.

"When I testify about health conditions in other states, they sometimes take steps. In Louisiana, they don't even listen."

Chemist Subra says that to make matters worse, refineries and petrochemical plants whose waste was labeled hazardous in the eighties applied to the DEQ to have the waste "de-listed," which would mean that they could dump it in solid and industrial-waste landfills at a cost two to four times less than in hazardous dump sites. "In the mid-nineties, I looked at de-listing applications from 100 firms, and all 100 were approved," she notes. The results of such actions were predictable: From

1995 to 1997 emissions jumped 7 percent (an additional 11 million pounds), a time when they dropped in most other states, says Templet.

Further, reprisals for whistleblowers are severe and certain. Templet says that to punish him for his activism at the DEQ, his promised salary at LSU was cut by $10,000. And according to Dr. Marise Gottlieb, a former researcher at Tulane's medical and public health institutions, she lost her research funding not long after she published findings about the connection between cancer and pollution in the *Journal of the National Cancer Institute* and the *American Journal of Epidemiology*. One article linked rectal cancer to drinking water from the Mississippi, while the other revealed the existence of clusters of lung cancer among those who lived within one mile of large petrochemical plants. When the legislature allotted Patricia Williams $1.1 million to hire physicians for a clinic to treat patients exposed to toxic chemicals, Governor Foster and Dr. Mervin Trail, chancellor of the LSU Medical Center, blocked the funds, and the clinic never opened. When asked why the clinic was torpedoed, Trail said the medical school had the right to spend its funds as it saw fit.

The Louisiana difference

Dr. Janette Sherman, an internist-toxicologist and author of *Chemical Exposure and Disease*, says that although Louisiana has no monopoly on pollution, the difference is in how its officials respond. "When I testify about health conditions in other states, they sometimes take steps. In Louisiana, they don't even listen," she says.

Templet compares Louisiana to many developing countries, in which a few industries and their political allies do well while most ordinary citizens do poorly. Based on 1990 census figures, roughly 30 percent of Louisianans live beneath or just above the poverty line, second only to Mississippi's 32 percent. Templet says that instead of educating and training the poor or creating jobs, the state gives massive subsidies to attract and retain industries—some $900 per resident in 1989, the year he studied this. Just as important, he insists that firms don't have to pollute to profit. Citing recent research, he says they could reduce emissions massively: In 1995 US chemical plants emitted 935 pounds of waste per job, while in Louisiana, the figure was 5,000 pounds. He contends that since companies manage to meet stiffer rules in other states, they could do the same in Louisiana. Elected officials argue that corporations would flee to greener pastures if the rules got too tough. But Templet disagrees: "They have everything they need here—the gas and oil, water transport and pipelines. And chemical firms spend just 1 percent or less of their total revenues on pollution controls."

Given the current climate, however, business will continue as usual. And Louisianans will continue to wonder whose child will be next.

4

Minority Communities Can Fight Polluters

Ziba Kashef

Ziba Kashef is the senior editor Essence.com, *an online magazine.*

The charge of "environmental racism" has been leveled at companies that build polluting industrial plants in lower-income, largely minority, communities. An important episode in this conflict took place in Convent, Louisiana, an African-American community where the operation of several heavy industries has brought about one of the nation's most polluted regions and the nickname of "Cancer Alley." In 1996, Convent resident Emelda West, a retired schoolteacher, decided to lead a protest against the building of yet another toxic plant in her neighborhood.

When Emelda West's neighbor called one July morning in 1996 to warn her about a new toxic chemical plant moving into their town of Convent, Louisiana, the then-72-year-old retired schoolteacher immediately thought about her daughter, Yolanda.

Mysteriously stricken with breast cancer at age 30, West's eldest daughter had died in 1982 after six years of suffering. "They said it was birth-control pills; I thought it was the environment," says West. Yolanda had lived in nearby Alsen, a Black community so heavily polluted by an incinerator and hazardous-waste landfill that residents organized to sue the waste-site owners to clean up their act. With her daughter's memory in mind, West didn't hesitate to agree that morning to join her neighbor at a community meeting to discuss the proposed plant, a large facility that would be located less than a mile from an elementary school.

West, a five-foot-tall powerhouse of a grandma with unwavering eyes, simply wasn't having it. For more than 40 years she had watched as Convent—a predominantly African-American town with little more than 2,000 mostly impoverished residents—gradually became crowded with one industrial plant after another. Foul-smelling odors from smokestacks and incinerators had long since replaced clean air, forcing folks to roll up car windows as they drove through the one-road-in, one-road-out town. West and her

Reprinted from "Saving Our Backyard," by Ziba Kashef, *Essence*, September 1999. Reprinted with permission from the author.

neighbors (many were descendants of slaves who had occupied the land) remembered the days when they could pull shrimp from the Mississippi River, and grow fig and peach trees for food. But after decades of development, West says, "I noticed the trees, just rotting and falling." Most concerned about the children of Convent, who frequently complained of nosebleeds, rashes and asthma, the mother of seven and self-described "education freak" notes: "They're absorbing all this poison."

So West and the mothers of St. James Parish (parish is Louisiana's equivalent of county) went to work, setting up meetings in their homes, local schools and churches to share details about the plant, known as Shintech, which would be producing polyvinyl chloride resin (PVC), a plastic used to make such products as pipes. In the manufacturing process the Japanese-owned facility would release several toxic substances, including a particularly dangerous chemical, dioxin, into the already toxin-choked air. Dioxin has been linked to health problems, including reproductive damage and cancer, according to published emissions data.

As West and the St. James group dug in their heels and organized against Shintech, other residents argued that the company would bring jobs.

The group's first task was to form an organization, St. James Citizens for Jobs and the Environment. Despite its initial efforts to block the plant at public hearings, both the local parish president and Louisiana governor Mike Foster welcomed Shintech. This was just the beginning of a David-and-Goliath battle that would pit the mothers and grandmothers against a multimillion-dollar multinational corporation. The St. James group had engaged in the larger war against "environmental racism"— the targeting of poor communities of color as dumping grounds for the nation's garbage and waste.

As West and the St. James group dug in their heels and organized against Shintech, other residents argued that the company would bring jobs. Shintech had promised to employ 2,000 workers to construct the plant and pledged also to practice affirmative action in hiring. In a town where a majority of Blacks were unemployed and those who had jobs earned an average of less than $5,000 a year, the prospect of such work was appealing. Even Ernest Johnson, the president of the Louisiana branch of the National Association for the Advancement of Colored People, signed an agreement with Shintech guaranteeing jobs to local residents.

However, West and other St. James Parish citizens pointed out that only 165 of those 2,000 positions would be permanent, and they feared most of those would not go to people in the community. "We've had 40 years of 'economic development' in St. James Parish," says Gloria Roberts, 72, the group's treasurer. "Yet people are living in deplorable conditions." Many African-Americans who applied for jobs at existing plants were told they were either underqualified or overqualified, forcing some to relocate in search of work. "Whether or not we educate our children in St. James

Parish, they're never good enough to acquire jobs in these industrial sites that are poisoning and killing all of us," says West.

The St. James Citizens continued to organize by circulating a petition—signed by more than 1,100 residents—and by writing letters to local newspapers. For legal and technical support and assistance, they also reached out to grassroots organizations, including the Louisiana Environmental Action Network (LEAN), Tulane Environmental Law Clinic and the Deep South Center for Environmental Justice. When St. James Citizens contacted Greenpeace—the well-known international environmental activist organization—their local struggle moved beyond Convent and became a symbol of the worldwide struggle for environmental civil rights.

If there was ever a clear case of environmental discrimination, Convent was it. Located along the Mississippi between Baton Rouge and New Orleans, the town was set in the heart of an industrial region dubbed Cancer Alley by environmentalists because of its elevated rate of cancer deaths. And St. James Parish had the third-highest level of toxic industrial emissions in the state, according to published data.

A survey conducted by the Deep South Center revealed a pattern of discrimination. "There was almost a perfect correlation between race and proximity to industrial sites," says the center's director, Beverly Wright, M.D. "The closer you are to the sites, the more Blacks you have." Robert Bullard agrees. He's the director of the environmental justice resource center at Clark-Atlanta University, which also provided support to the people of Convent. "You have this historical pattern, the population being African-American and poor and all these plants concentrated in one area," he notes. "It was naked racism."

So on behalf of St. James Citizens, on April 2, 1997, Greenpeace and Tulane University Environmental Law Clinic filed the first of several complaints with the Environmental Protection Agency (EPA). The first grievance objected to air permits issued to Shintech by the Louisiana Department of Environmental Quality (LDEQ). Additional complaints charged discrimination: They cited President Bill Clinton's executive order that requires state agencies funded by the federal government to consider the impact of pollution on poor and minority communities. "The Shintech case emerged as the most important civil-rights case ever involving charges of environmental racism," says Damu Smith, the southern regional representative for Greenpeace. "Other communities would be affected by whatever decisions were made."

The legal maneuvers worked, at least temporarily. On September 10, 1997, EPA president Carol Browner stalled Shintech's air permit on technical grounds. In a letter to the LDEQ, Browner asked the agency to hold public hearings on the plant and consider the impact on the predominantly Black community. She wrote:

"We believe it is essential that minority and low-income communities not be disproportionately subjected to environmental hazards." For the first time in EPA history, the agency also launched its own investigation into whether the LDEQ violated Convent citizens' civil rights.

To put further pressure on Shintech, in April 1998 West and Roberts went to Washington, D.C., where they met individually with EPA representatives and congressional Black Caucus leaders, including Maxine Waters,

Jesse Jackson, Jr., John Conyers and Bobby Scott. The two septuagenarian "lobbyists" also held a press conference in D.C. After Greenpeace's Smith and Minnesota Senator Paul Wellstone addressed the reporters, Roberts used a map of the parish to point out the 16 toxic facilities in the area, and talked about how Shintech would dump 600 million gallons of waste material into the Mississippi each year. When West took the podium, she spoke without a script. "I have a lot of pent-up emotion," she began in a trembling but searing voice. "We have our two schools, which are 95 percent African-American, our Head Start and our elementary school on each side of the River. Those are my kids. They're trying to kill our future generations."

In June 1998, West and Greenpeace representatives went to Japan to press their case in person at Shintech's parent company, Shin-Etsu in Tokyo. Company representatives were reluctant to meet with them, but Smith called every day: "I told them I have a 72-year-old African-American grandmother who wants to meet with you. You cannot refuse a meeting while we are here in your country." After several days, an assistant to Chihiro Kanagawa, Shin-Etsu's CEO, agreed to receive the petitions and letters from Convent residents that West had brought with her. Despite the cordial but cool reception from Shintech executives during an hour-long meeting at the company's corporate headquarters, West ended her presentation by pounding her fist on the lectern and crying out,"We don't want Shintech!"

"I have a lot of pent-up emotion. . . . We have our two schools, which are 95 percent African-American, our Head Start and our elementary school on each side of the River. Those are my kids. They're trying to kill our future generations."

After 18 months of work, West, Roberts and St. James Citizens finally slew their Goliath. On September 17, 1998, Shintech announced it would suspend efforts to obtain permits for the Convent plant and proposed building a smaller facility in the nearby community of Plaquemine. "The state of Louisiana really wants that industry here," says West. So instead of celebrating, "we should just have a thanksgiving and praise the Lord and wait." Now galvanized by her own fight, West has lent her support to her neighbors in Plaquemine—a racially diverse middle-class community—in their quest to stop Shintech.

The women of Convent have used their fame to educate others. "The Sierra Club came to listen to our story," says Roberts. "The Rockefeller Foundation president came and we took him on a tour of the area where people are being impacted most by the toxic air."

With her deceased daughter and Convent's schoolkids as her focus, West explains why she won't give up: "There's a danger with industry. You can't walk away from it now." After a pause, she adds, "The fight isn't near over."

5

Communities Will Tolerate Pollution When Jobs Are at Stake

Bobbie Ann Mason

Bobbie Ann Mason is a novelist and short-story writer who was raised in western Kentucky.

From its founding in 1952, the Paducah Gaseous Diffusion Plant was a good neighbor to its employees and to the surrounding community of Paducah, a small city in the traditionally agrarian region of western Kentucky. But the radioactive isotopes created while the plant produced enriched uranium for weapons as well as peaceful energy production have brought about illness and death among workers, most of whom saw themselves as taking part in an important and patriotic enterprise. Attitudes among many local workers haven't changed, even as the news broke in the national media that Paducah has been suffering through one of the longest and worst toxic waste-related health crises in the nation's history.

On the national radar screen, Paducah, Kentucky, is a provincial town with a funny name, but here in the western end of the state it was never an inconsequential place. I grew up on a farm near the small town of Mayfield, and Paducah was so far away—twenty-six miles—that we went there only on special occasions. It was the city, the Mecca for several counties of farmland. It had department stores, fine ladies' shops, movie theatres. I was dazzled when, as a child in the nineteen-forties, I went shopping with my grandparents on a Saturday. We dressed up and wandered through the riverfront Market House, where exotic produce—even oysters—arrived by train. On the way into Paducah, we passed the railroad repair shop, with huge locomotives squatting in the yard—an impressive sight that made me think important industry occurred here, something that linked our area to the whole world.

Downtown Paducah was ritzy. In high school, I attended a dance at the swank pseudo-Tudor Hotel Irvin Cobb, named for a novelist and humorist

Reprinted from "Fallout," by Bobbie Ann Mason, *The New Yorker*, January 10, 2000. Copyright © 2000 by Bobbie Ann Mason. Reprinted with permission from International Creative Management, Inc.

who had appeared in a movie. He won the 0. Henry prize for the best short story in America in 1922. But Paducah's true star was Alben W. Barkley, who was born in a log house and worked his way up in society and politics through a long career in Congress, eventually becoming Vice-President under Harry S. Truman. After President Truman dropped atomic bombs on Japan to end the Second World War, Barkley—Paducah's favorite son—contrived to bring one of the nation's first atomic plants to his home turf. Everybody called it the bomb plant, even though it didn't really make bombs; it processed and enriched the fuel for them. Paducah, Barkley argued, had just the location—the site of the Kentucky Ordnance Works, which had manufactured explosives for conventional bombs during the last war. It was a logical shift to the new technology that America's defense depended on. And, in a gesture of pork-barrel politics gone nuclear, Barkley bequeathed a lasting gift to his hometown—uranium.

Doomsday wasn't going to happen in Paducah, not as long as the plant helped supply the nation with its friendly nuclear arsenal.

The Paducah Gaseous Diffusion Plant began enriching uranium fuel in 1952. This top-secret business was something like increasing the octane in gasoline—putting the oomph in the bomb. Helping to create A-bombs was a giddy, lucrative endeavor, and Paducah began to change. The first apartment buildings I ever saw shot up while the plant was being built, in the early fifties. Twenty thousand construction workers jammed the town, and local people rented out spare rooms and barn stalls—any available nook—for the newcomers to sleep in. At a time when "The Walt Disney Story of Our Friend the Atom" was a popular book, and [Soviet premier] Nikita Khrushchev's name was on every tongue, and Sputnik was terrifyingly in orbit above us, Paducah was called upon to be alert—and secretive—in exchange for good jobs and the chance to beat the Russians. It wasn't afraid. Doomsday wasn't going to happen in Paducah, not as long as the plant helped supply the nation with its friendly atomic arsenal. Today, the plant enriches uranium for nuclear-power reactors instead of atomic bombs, but workers still affectionately call it the bomb plant. License-plate holders from the nineteen-seventies show Paducah's namesake, Chief Paduke, on the left and an atomic cloud on the right, framing the words "Paducah, the Atomic City."

Greensalt and toxic trash

Last summer, Paducah's deal with atomic energy seemed to be exposed as a bargain with the Devil. The news was packaged in one explosive bundle in the Washington *Post,* on August 8th: radioactive-waste dumps, safety violations, bureaucratic lies, cancer, environmental pollution. Whistle-blowers, in a sealed lawsuit filed in June (it has since been opened), charged that former operators had defrauded the government by covering up knowledge of widespread radiation contamination, without regard for the safety of the workers. The *Post* reported stories of nuclear waste being treated lackadaisically—as if it were no more dangerous than kitchen compost. Workers routinely breathed heavy black uranium dust, and some said that supervisors sprinkled it on the cafeteria food—to prove that the dust was harmless. Workers handled so much of a uranium compound called green-salt that their skin turned green, like the Jolly Green Giant's. Their bed-sheets at home were stained green.

Even worse, the *Post* reported that many workers had also unwittingly handled plutonium—for decades. Plutonium, which is deadly and cancer-causing, was never supposed to be at the Paducah plant. It had arrived during the Cold War, along with other highly radioactive fission by-products, as an impurity in shipments of used uranium. And it remained in the plant, like an unwelcome guest, dirtying up the place. The plutonium was in the uranium dust that the workers breathed.

The workers never made much fuss about safety conditions, although the plant was becoming its own toxic-waste dump—tons of radioactive scrap metal and cylinders of depleted uranium were piling up with nowhere to go. Toxic trash was tossed over the fence into an adjoining wildlife area, and local wells had become contaminated. Then plutonium was detected in a ditch outside the fence. And a radioactive, technetium-tainted underground plume of water was inching toward the Ohio River. Paducah, exemplary heartland town, where people went to church and gave the time of day to strangers, thought it had been spared such modern ills. The local press—the Paducah *Sun*—had downplayed the dangers. Until the Washington *Post* showed up, it was as though all the toxic trash were just part of the furniture—the price Paducah paid to have a thriving economy, the price paid to help win the Cold War.

"The plant gave people good jobs. It kept a lot of people from starving. And now look what's happened. There's so much good—and so much horror."

Plutonium is heavy and it doesn't move fast, but, when I heard that it was present at our local nuclear-fuel refinery, I felt as if a plutonium-polluted plume were headed toward me. This wasn't Chernobyl—a nuclear power plant run amok. It was personal. My sister had worked at the bomb plant for several years during the late seventies. I E-mailed her in Florida, where she now lives. "I guess I was exposed," she answered. "But don't worry. If you got it you got it, and there is nothing that can be done—but maybe it can for the next generation." She reminded me how good the plant had been to its employees. The salaries were the highest around, and the benefits were off the scale. Besides, she told me, everything was so secret there. Nobody talked. You felt you were doing something important, something good for the country.

It was late in the dust-bowl summer, and dust from the desiccated fields sifted onto my car when I drove to Paducah from my home in central Kentucky. I was trying to think of a reasonable synonym for "freaked out," which was just about how I felt. The guy who pumped my gas said there was too much else to worry about in this world for him to be concerned about loose isotopes or technetium seepage. Wars, earthquakes, and such.

From Paducah, I continued west to Future City, where construction workers were housed when the plant was being built. Now it was just a crossroads, with a grocery and a barbecue eatery. The bomb plant was nearby, and just up a parallel road was Heath High School. The school had been the site, in 1997, of one of the first of the string of school shootings. A fourteen-year-old boy gunned down three of his classmates in a prayer meeting in the lobby. I remembered the reporters and the television crews descending on the place like paratroopers, and I imagined that Paducah must feel jinxed now. The atomic-waste scare was bringing the news crews rushing back. The same barbecue joint fed both frenzies. I came to a halt at the crossroads. I was reluctant to look at either of these scenes—the plant or the school—and I didn't know if it was from fear or from sadness. I drove back to Paducah.

An age of innocence

Dottie Barkley has been a family friend for twenty years. When she was seven years old, she was a guest on "I've Got a Secret," one of the old TV quiz-game shows. Her secret was that her grandfather was Vice-President Alben W. Barkley, Paducah's local hero. On another occasion, she was taken to the Plaza Hotel in New York, where she pitched a fit because the Palm Court didn't serve catfish. She cried, "If I can't have catfish, then I don't want anything!"

We were sitting in Dottie's back yard among morning glories and gourd vines. Seven cats and Winnie, a chow-shar-pei mix, crowded around us. "This is outrageous," Dottie said, fuming about the local coverage. "The other Sunday, the Louisville *Courier-Journal* had a big front-page story about plutonium in Paducah, but there wasn't a peep about it in the Paducah *Sun* until the next day. Most of Paducah didn't even know about this!"

Did she think people felt betrayed? I asked.

"Hell, most people don't really care," she said. "Everybody at the plant knew they were working with dangerous stuff. Maybe they didn't all know it was plutonium, but they knew. Now people don't want to talk about it. They don't want to lose their jobs."

To visualize Dottie, imagine Marilyn Monroe—outfitted by the Limited—with a pickup truck. She is glamorously bohemian, with a blond heap of curly hair. And, as Alben Barkley's granddaughter, she holds a unique social position in Paducah, even though she avoids the cocktail-party carrousel of the local big dudes, where she might rightfully belong—not her style. She works at the Party Mart ("Paducah's Most Interesting Store").

Twelve ounces [of plutonium] is enough to provide a maximum legal limit of ingested plutonium for about two and a half billion people, or nearly half the world.

She explained how the plant got here. "Granddaddy just muscled it through. He was best friends with Speaker of the House Sam Rayburn. And it was such an exciting thing when the plant came! The plant gave people good jobs. It kept a lot of people from starving. And now look what's happened. There's so much good—and so much horror." She shuddered. "Granddaddy couldn't have imagined this. He couldn't have known how it would turn out. Could he?"

A cat named Dinah Shore jumped into my lap. "Dear Hearts and Gentle People"—the song by the real Dinah—ran through my mind. I thought of the fifties, when a war-weary nation quelled its fears of the bomb by listening to songs like this, or to Doris Day's "Que Será, Será."

Dottie said, "The people out at the plant were so innocent back when it started. They used to handle uranium with their bare hands!" She bent over briefly to hug her dog. "You know, you should talk to my ex-husband Joe. He's worked at the bomb plant for twenty-six years, and he can tell you what it's like. They're like a big family out there, and he's been exposed to just about everything."

As I was leaving, I noticed a photograph in the dining room of Dottie with her parents and her grandfather, taken when Barkley Airport was dedicated in 1948. Dottie, a child in a shiny taffeta dress with a cross-sash, has a missing front tooth.

"Look at me in my queen outfit," Dottie said, laughing at herself in the picture. "Get Joe to take you out to the bomb plant. It's so eerie."

A different Joe, Joe Harding, had known about the dangers years ago. He died in 1980 of stomach cancer. He began working at the plant in 1952, and his jobs included flushing impurities out of the processing pipes. Apparently, a residue of all kinds of radioactive things—plutonium, neptunium, and other contaminants—remained in the system once the processing was completed. Even today, there is a residue clinging to the pipes, like what's left in the skillet after you cook onions. Harding was chronically ill, but when he declared that he had radiation poisoning no one believed him. He had weird toenail-like growths coming out of his elbows and kneecaps, but people only laughed. The management said his illnesses were caused by eating too much country ham. His disability benefits and insurance claims were denied. A few years after his death, lawyers representing his widow ordered that his body be exhumed, and his bones revealed a level of uranium hundreds of times above normal.

Normal? Radiation is good for you; it boosts the immune system, according to a local engineer in an August 23, 1999, *Sun* column. "Radiation . . . may benefit the health of those exposed . . . a low dose of radiation actually increases immunity," he wrote. The plant's neighbors must glow with good health, then, because radioactive technetium-99 has turned up in the gardens—in banana peppers and turnip greens. Traces of plutonium were found in deer in the wildlife sanctuary—not enough to hurt you, officials said. To be in any kind of danger, "You would have to eat the whole deer," the Kentucky state health commissioner insisted on TV last summer—his remark delivered with the fervor of a political stump speech. Why, I wondered, do people always seem to be telling us that we can eat radioactive waste?

Acceptable risk

Downtown is "Historic Paducah"—antique stores, funky shops, and Saturday-night street parties. Like many other towns, Paducah is energetically reclaiming itself from the mall, and you can almost imagine the main street in its heyday. Tourists from the Mississippi Queen and the Delta Queen stream in through gates in the flood wall, which is being painted with murals depicting the history of the city. The showpiece of downtown is a quilt museum. Paducah is morphing from the Atomic City to the Quilt City. You might think that quaint old quilts are a clever atomic-age coverup, but the museum is on the cutting edge. Its quilts are postmodern.

I was headed for the Paducah Public Library. I had been mulling over the phrases "acceptable risk" and "eat the whole deer." ("I can't believe I ate the whole deer!") And, for that matter, just what is a "trace"? What is an acceptable number of picocuries of plutonium? How many would you want to have settle in your brain, your lungs, your islets of Langerhans?

At one time, five hundred picocuries of plutonium were detected on the plant grounds—thirty-three times what the government deemed an acceptable standard at blast sites in the South Pacific.

Plant managers claim that the amount of plutonium that came to Paducah was only twelve ounces—a piddling amount. Neptunium may be a worse problem. It is less radioactive than plutonium, but forty pounds of it were brought to the plant in the ill-fated uranium shipments. Plutonium and neptunium are transuranics, metals that are heavier than uranium. They are artificially created radioactive elements. They don't occur in nature; they pop up when atoms are split. I knew that plutonium is a hundred thousand times as radioactive as uranium, with a half-life of twenty-four thousand years—longer than civilization has existed. A beeline to the encyclopedia revealed to me what no one was admitting: twelve ounces is a lot. Theoretically, that much plutonium contains as much energy as nearly six thousand tons of TNT. More to the point, it's incredibly toxic, even in microscopic amounts. The "safe" dose for a human being is 0.13 micrograms. Thus twelve ounces is enough to provide a maximum legal limit of ingested plutonium for about two and a half billion people, or nearly half the world.

Around town, though, people didn't seem worried. There were virtually no letters to the editor of the Paducah *Sun*, and few people seemed willing to voice any fear of atomic pollution—as though talking about radioactivity might be enough to shut the plant down. TV NewsChannel 6 (which is owned by the *Sun's* parent company) seemed more alarmed by weather scares than by radioactive-waste dumps or the presence of plutonium in the food chain.

I chatted a while with Iris Garrott, one of the librarians, and she reminded me that that was how folks were around here. "There's a sense that they took the risk for the jobs," she said. "They went along with it. People here are concerned about personal and emotional things—like the shooting at Heath. That's when everybody gets in a stir, when it touches you personally." Iris, who had three flashy earrings in her left ear and one in her right, leaned forward. She said, "But the news is sinking in. Every day, something new comes out. Everybody's on edge, I think. They're just waiting."

For what?

"They're waiting for somebody big to come to town—Energy Secretary Bill Richardson, Tom Brokaw. If Tom Brokaw came, then it would be real."

Going on a treasure hunt

Joe Gorline, Dottie's ex-husband, loves working at the bomb plant, as his father did. From the start, Joe's father told him, "This stuff is not good," and Joe has been careful. His father died from chlorine-damaged lungs, but Joe has been loyal to the plant and to its important secret work.

He looks strong and healthy. He is tan and muscular, somewhat large in the middle, and has a long gray ponytail, fastened at intervals with colored rubber bands. He lives with his pit bull-Rottweiler, Baba Ram Dass (Bubba, for short). In his house are a Finnish 20-mm. antitank rifle and a Second World War German MG 34, and a safe filled with his gun collection.

Joe repairs equipment in what's called the cascade—a six-hundred-mile complex of pipes which comprises the enrichment system. Uranium hexafluoride gas, or UF_6, is sieved repeatedly in the cascade to get a richer concentration of uranium—a panning-for-gold procedure. Joe might replace joint seals or weld pipes that carry UF_6. For such work, he wears a safety suit with a respirator. "I call it my banana suit," he told me. "It's yellow, with yellow rubber boots, orange gloves."

Orange gloves? I asked.

"It's a statement," he said with a grin. "Accessories are everything."

Doesn't he get hot in that suit?

"Oh, I'm used to the heat. But it's noisy! It's like being in the crankcase of your car. I haven't heard a bird chirp in years." He laughed and cupped his ear. I listen to the machines. My job is to keep things running. After a while, you want it to run. You develop pride in your work."

Wasn't he afraid of radioactivity?

"There's nothing out there now that scares me. The safety has improved. But if anybody got big doses I did. Fifteen years ago, the same place I'd go now in my banana suit, I went in with coveralls. We'd get covered with black oxide dust and it would be all over us, and then we'd go to the cafeteria and wallow around. I got UF_6 and black oxide in my mouth and eyes." He laughed. "It tastes terrible!"

He showed me a small crater on the side of his nose, where he was burned by fluorine. "It would condense on a vent and drip over the door. One day, a guy went out the door and something dripped on his nose. He went nuts. He thought he'd been burned until we yelled, 'No, that's just pigeon shit!' But it dripped on me one day. My nose started smoking on the end. When you start smoking, you have to go to the dispensary. One guy's nose was worse than mine. It was flat."

I knew I ought to see the plant, but I wasn't sure I wanted to go. I'm health-conscious. I consume antioxidants, count fat molecules, pick organic turnip greens. Did I really need to go on a treasure hunt for transuranics? Should I carry a Geiger counter? Wear throwaway shoes?

I rode to the plant with Joe in his old Chevrolet truck. He flicked his cigarette discreetly out the cracked-open window. Paducah's urban sprawl is westward, toward the plant. Around the mall are the typical clusters of ugliness which deface America. In the subdivisions, Ten Commandments signs had sprouted in people's yards. The tall houses in the extravagant new developments looked overpriced and too close together, but I was glad that at least they weren't usurping all the fine farmland. Just past Future City, we turned onto a road that led to the plant, lined with small farms and modest homes. The corn was dying.

"The soybeans are strangely green," I said.

"It's plutonium that does it," Joe said. "All that radiation."

Sometimes you have to take what Joe says with a grain of greensalt.

The plant, which occupies a fenced-in area nearly the size of Central Park, is a sprawling gray complex. The architecture resembles the back end of a shopping mall. Right away, I noticed what was stored in the front yard—the blue cylinders of depleted uranium, rows and rows of them. Each of these cylinders—there are thirty-seven thousand of them— weighs between ten and fourteen tons. They made me think of a stockpile of pods from "Invasion of the Body Snatchers." The cylinders will be there until someone figures out an economical way to recover the last traces of valuable uranium in them. In the meantime, everyone hopes that they won't rust, or leak—or explode.

As we drove around the outskirts of the plant, I glimpsed some of the "hot spots" the Washington *Post* had written about—small areas where toxic waste had been spilled or buried or dumped. They were roped off and low to the ground, with little warning signs. At one time, five hundred picocuries of plutonium were detected on the plant grounds—thirty-three times what the government deemed an acceptable standard at blast sites in the South Pacific. Some of the buildings are so heavily contaminated that they have been abandoned, and, unfortunately, wildlife now live in them.

The plant itself is not a reassuring image: it's aging and corroding. There are six processing buildings, and they are all hooked together with pipes—long, unsupported, seemingly precarious overhead pipes. I was trying to grasp the way the plant worked, the way the gas was pumped through a network of compressors and converters—the cascade. It was such a mysterious concept—waterfalls, something beautifully flowing— that for a moment I almost wanted to see it. My sister saw it once when she worked in the safety department. A supervisor took her inside one of the big buildings which housed part of the cascade. He wanted to show her what their work was all about.

We turned north, on a gravel road alongside the chain-link fence, and spectacular waste dumps came into view: the rusty scrap heap, the old cylinders, the giant mound of crushed fifty-five-gallon drums (Drum Mountain, it's called). Uranium, radionuclides, "uranium daughters" (a phrase that captured my fancy), and transuranics infused these collections like mildew in damp clothes. Neptunium, plutonium, technetium, old kitchen sinks—it all seemed to be here in the scrap piles, as common-looking as a junk yard of wrecked cars. In a way, the scene seemed normal. It's a time-honored rural practice to save your trash in the yard— what won't fit on the porch. You dam the creeks with old mattress springs and broken refrigerators, to stop soil erosion. An earthquake on the dreaded New Madrid fault could turn this region, embraced by the Ohio, Mississippi, Tennessee, and Cumberland Rivers, into mush. Where would all this irradiated trash wind up then?

The plant had been built on a thirty-four-hundred-acre federal property, where the old munitions plant—the original Kentucky Ordnance Works, or the K.O.W., as it was called—operated during the Second World War. Most of the land that surrounds the plant is the wildlife area and extends almost to the Ohio River. Joe was driving through this wilderness now. I knew that there were still toxic chemicals from the ordnance works in the ground. (The K.O.W. had manufactured TNT.) And now radioactive pollution had spread through this area. The Washington *Post* said that two dozen radioactive rubble piles from the bomb plant dotted the landscape, but I didn't see any. Radioactivity was an invisible, ghostly presence.

"Over there's a great pond for frogs," Joe said. "I used to frog gig there a lot with my son."

Neighborliness, not litigiousness, has always been the norm around here, and the social contract meant getting along by going along.

We passed other ponds, where recently the fish had been killed so that nobody would eat them. People have hunted and fished here for decades; no one wants to throw back a good catch. I was aware that this wildlife area is virtually sacred. People feel so deeply about hunting here that they would be up in arms, so to speak, if the area were condemned because of mere toxic waste. We twisted and turned down gravel roads; then we were in the scrubby fields, driving on what were just old worn paths that Joe said he knew by heart.

"I love this old truck!" he burst out gleefully as we bounced over a bump. The tracks ran beneath thick weeds and tall grass. We were in a labyrinth of ancient trails. In a clearing, we passed a group of teen-agers slouching around a pickup truck, playing hooky. This park is where kids come to party, Joe told me, and schoolchildren have picnics in these fields, where, over the years, thirty tons of uranium were flushed into the streams, saturating the earth, and recently an unmarked pile of contaminated railroad cross ties was discovered.

We passed a pair of sirens on poles, with signs—what to do if the siren sounds. (Basically, run for at least two miles.) A concrete water tower loomed ahead, then another and another. We were among the ruins of the ordnance works—concrete hulks. Vines crawled over the gray shapes. Even though nature was taking over, the landscape itself was a ruin, shrivelled by the drought, the sumac and sassafras reddening prematurely.

I had lost my sense of direction, and I didn't fancy crawling through scrubland fertilized by uranium or TNT, but Joe wasn't worried. It was a hot day. He had a jug of water and a cooler of Cokes. I clutched my bottle of water from France. We got out of the truck and waded through tickseed and ironweed, then down a gravel path. Joe, sockless in sandals, reminisced about youthful outings here as we peered inside some of the dank old buildings. They were dark, with graffiti-covered walls. A disintegrating couch, its stuffing spewing, sat beneath some lingering asbestos that hung from the ceiling like Spanish moss.

I was either in a gothic romance novel or in an apocalyptic Italian movie. I faced a wall, spray-paintcd with a message: "Live in Fear, the End Is Near."

I felt suddenly uncomfortable to be in a place that had an unhealthy obsession with bombs and guns and other insidious things that kill people. I saw the K.O.W. as the ancestor of the bomb plant, and I knew the plant was creating its own ineradicable legacy. The sins of the past—uranium daughters—lay strewn over the landscape and in the water and under the earth. From TNT-based weaponry in the Second World War to the first atomic bombs and the nuclear stockpiles of the Cold War, the wartime urgency left a habit of mind and a profusion of poisons.

Dottie had shown me a photograph Joe took of her out here. She was standing in an open window in one of these ruined buildings, and in the photo the light had created an apparition above her head, merging with her bright hair. She said it looked like a dove, but its glaring whiteness reminded me of an atomic blast.

On another day, screwing up my courage, I returned to the plant alone, determined to see the cascade for myself—heat and noise and all. It was a bleak, gray, rainy day, but after the drought I was glad to see it. I was in my rain gear, with bright-yellow boots. I wished I had Joe's banana suit.

My guide wore a thermoluminescent dosimeter, a radiation-monitoring badge, but she said that I didn't need one, since I wouldn't be allowed anywhere in the plant where there might be radiation. I was a little disappointed but mostly relieved. I was allowed into the control center, a round domed concrete building. Inside, on the curved wall, was an immense diagram of pipes and compressors and converters and electric motors. It was fifties technology, intricate but decidedly pre-Microsoft. It was like the cockpit of Captain Video's spaceship. The diagram on the wall, with lots of red and green lights and dials, mimicked the cascade. The whole system has never been shut down since it started, in 1952. If it were shut down, the gas would cool and turn into a solid, and the cascade would clog up like a cholesterol-choked artery. A gauge on the wall—like a big clock—had a dial indicating the gas level.

I was left to imagine the mighty cascade. It was like a Rube Goldberg cartoon version of the human circulatory system—the crudest technology for something as mysterious as a beating heart. The heart of the mystery of atomic energy, its deadly magic, was a mundane industrial process. Somehow, I could picture Lucy and Ethel in here running this thing.

Neighborliness

The onset of autumn brought a startling revelation: an accidental uncontrolled nuclear chain reaction was theoretically possible in the plant. Paducah jumped out of its time warp, crashing into the twenty-first century. People were confused and scared. Energy Secretary Richardson visited and promised the moon. He apologized for the plutonium. The plant was buzzing with investigators. Joe E-mailed me, "During the day the plant is a hotbed of activity, auditors everywhere. They don't know whether to shit or go blind." In September, a ten-billion-dollar class-action lawsuit was filed against former contractors, including Union Carbide and Lockheed Martin, claiming, among other things, mental distress and "unjust enrichment." Joe wouldn't join it, and he had no kind words for whistle-blowers. "If I get cancer, I don't even want to know," he told me.

I've been trying to put my finger on why, for so long, Paducah remained passive in the face of danger, something I feel I know intuitively as an insider but which seems to befuddle outsiders. Why did the workers trust some of the government contractors that ran the plant? Did they really believe that giant corporations would look out for their well-being? How could they have been so innocent? Is that how those contractors got away with their colossal abuses?

These are post-Vietnam questions. The same people who are asking these questions seem a bit wistful about the virtues of small towns. All I can say is that such things exist. People here haven't yet plunged into the frantic greed frenzy of the big time. They're independent, proud people— agrarian, basically. They don't want to be told what to do—like "don't hunt on the wildlife refuge"—but once a bargain is made and a trust is built, as it was with the plant from the beginning, they will honor it and they will do as they are asked. It was more than high-paying jobs. Neighborliness, not litigiousness, has always been the norm around here, and the social contract meant getting along by going along. The problem was that the plant had been a good neighbor. It was good to its workers, who kept the secret well.

I'm drawn again to Future City. At the intersection, it seems that the future is nuclear fallout in one direction and guns in the schools in the other. I turn north, toward Heath High School. Driving up, I see kids across the road at band practice. A banner outside the entrance of the school reads, "Rising to the Challenge." Inside, in the lobby, fourteen-year-old Michael Carneal opened fire on the prayer group two years ago. The Paducah Gaseous Diffusion Plant—a good neighbor—contributed generously to the memorial fund for the three dead girls.

I recall that, for some of the students, the instinctive reaction to the massacre was forgiveness. They painted banners that said "We forgive you, Michael" and "We love you, Mike."

And that's the heart of the story. This turning the other cheek, the strange embrace of sudden horror, startled outsiders. The students' anger came later, when grief had set in and the lawyers showed up, but their initial acceptance—their passive non-resistance—was not so surprising in an agricultural region, where farmers forgive the forces they cannot control. Droughts and pestilence are risks the farmer takes at every planting time in every hopeful spring.

6

The Chemical Industry Seeks to Keep Vital Information from the Public

Claude Morgan

Claude Morgan is a freelance writer living in Portland, Maine.

By the Clean Air Act of 1990, the United States Congress required an estimated 66,000 U.S. chemical facilities to make public a record of their inventories, their accident histories, and "worst-case scenarios" in the event of a fire, a spill, an explosion, a mechanical failure, or other incident that might expose the public to a disastrous release of toxins into the environment. The information was to be submitted to the public and published on the Internet in 1999. But the chemical industry is fighting the electronic publication of worst-case scenarios or accident histories, claiming that terrorists or criminals can use this information to blackmail, threaten, or sabotage chemical factories. Helped by their allies in the federal legislature, the chemical companies are seeking to avoid embarrassing disclosures about their safety records and to forestall any public accountability for their actions.

Information can be hazardous to your health or so says the chemical industry.

The chemical industry is attempting to dilute a law that guarantees the public's right to know about what could happen to communities surrounding chemical plants in the event of a serious accident.

Congress passed a community right-to-know provision as part of the Clean Air Act in 1990. The law has required an estimated 66,000 U.S. chemical facilities to compile a record of their inventories, accident histories, and worst-case scenarios. The companies were supposed to submit this information to federal authorities [in 1999].

To comply with the right-to-know law, the EPA planned to make the information public, including putting it on the Internet. But that's when

Reprinted from "Warning: Information Can Be Hazardous to Your Health—or So Says the Chemical Industry," by Claude Morgan, *The Progressive*, August 1999. Reprinted with permission from *The Progressive*.

the chemical companies balked. They claim the risk of Internet terrorism is greater than the risk of chemical disaster.

Banning information

In May [1999], Representative Tom Bliley, Republican of Virginia, introduced legislation that would prohibit government employees from publishing the industry's worst-case scenarios on the Internet or in any "electronic form." The law would also bar police, firefighters, local emergency planners, and federal, state, and local officials from discussing chemical hazards with their communities. And in language that many critics say undermines the 1966 Freedom of Information Act and violates civil liberties, Bliley's bill would require librarians to track and monitor library users who request information about the dangers of chemical industry accidents.

Bliley is the chairman of the House Commerce Committee and a longtime champion of the chemical industry. According to the Federal Elections Commission web site, he received $55,000 from the chemical and allied products political action committee in 1998.

Uses of the Internet

At an April [1999] computer security conference in Washington, D.C., chemical industry advocates warned that only two groups would benefit from putting the worst-case scenarios on the Internet: terrorists and environmentalists.

"On the one hand, posting this information over the Internet makes it easier for folks who want to describe the magnitude of the problem to the nation," cautioned Jamie Conrad, legal counsel for the Chemical Manufacturers Association. "The other people that it makes it easier on is criminals and terrorists who'd like to blow things up and make a big bang."

"Seventeen million people around the globe have the skills to launch a cyber-attack."

Another conference panelist darkened the dire picture. "Seventeen million people around the globe have the skills to launch a cyber-attack," said Jody Westby, Internet consultant and former policy analyst for the Progress and Freedom Foundation, a conservative think tank monitoring technology and its effect on public policy. "The real threat is that they'll take this information and target facilities and sit in Iran or wherever they want to be and cause a worst-case scenario through an information-technology system. I challenge any of you to look at this form, and then say you'd like to have Mr. bin Laden [suspected terrorist Osama bin Laden] have it."

But the conference attendees weren't buying the industry's message. Posting worst-case scenarios on a web page doesn't provide terrorists with an entrance point to chemical facilities or their computer systems, said Simson Garfinkle, chief technology officer for Sandstorm, Inc., a Cambridge, Massachusetts, security-software firm. "You're just confusing people," he told chemical industry representatives on the panel. "You're trying to scare us."

The community's right to know

"This is about companies not wanting to be embarrassed over the accident potential they create," says Timothy Gablehouse, a Denver, Colorado, lawyer and chair of the Jefferson County local emergency planning committee. "It is frankly unbelievable that Congress would attempt to restrict my speech on these topics," he testified before the House Commerce Committee in May [1999].

"It is not possible to have a meaningful conversation with community groups about how to protect themselves if we do not discuss the accident scenarios they may face."

"Community-right-to-know has greatly helped to reduce chemical hazards, to get companies to take notice of, and take responsibility for, community safety," says Paul Orum, coordinator of the Working Group on Community Right-to-Know, a nonprofit organization that tracks government and industry compliance with community right-to-know laws nationwide.

"It's time for industry to stop lobbying for secrecy," says Orum. "If chemical companies want to promote this as a security problem, then they should address it as a security problem."

"We can't find any meaningful studies to suggest that chemical facilities are at a greater risk to terrorism now than they have ever been before," says Rick Blum, public affairs liaison for OMB Watch, a public interest group focusing on access to information at the federal Office of Management and Budget.

The security study

Blum says that, so far, he and his colleagues have identified only one study that addresses the chemical industry's vulnerability to terrorism. It is Security Study: Analysis of Terrorist Risk Associated With the Public Availability of Offsite Consequence Analysis Data under EPA's Risk Management Program Regulations. Commissioned by the EPA from the Aegis Corp. and ICF Incorporated in 1997, the report says that posting industry's worst-case scenarios over the Internet could double a facility's risk of terrorist attack. Members of the chemical industry cite this as proof that worst-case scenarios should be barred from the Internet.

In 1997 alone . . . U.S. chemical facilities reported 38,305 chemical accidents to the EPA—roughly one chemical fire, spill, or explosion every fifteen minutes . . .

That study, however, is riddled with flaws, Blum says.

"Because of the lack of historical data and the limitations on existing risk assessment methodology," says Kathy Jones, EPA's associate director for its Program Implementations and Coordination division, "the report did not provide estimates of the absolute levels of risk associated with a terrorist attack on a chemical facility." Instead, says Jones, "it focused on the relative levels of risk."

For starters, says Blum, the report fails to establish a clear baseline of threat—which might easily be "something close to zero" because there has been no terrorist attack. "Basically, if you were to ask the authors of this study if posting a baseball game over the Internet would be a threat," says Blum, "they'd have to say 'yes, because you'd be telling terrorists where to find 50,000 people gathered in one place." The study's director has denounced many of the conclusions that lawmakers and industry are now drawing from it. "I am given to understand that the relative risk projections in the Security Study have been interpreted by some parties as a reason to question the merit of EPA's plan to make the Risk Management Plan data available on the Internet," wrote Howard Dugoff, vice president of ICF and the director of the study, in a letter to James Makris of the EPA. "I believe this interpretation is grossly inappropriate."

Three members of Congress (Sherrod Brown, Democrat of Ohio; John Dingell, Democrat of Michigan; and Ron Klink, Democrat of Pennsylvania) have called for a federal agency to investigate the study's methods. A Government Accounting Office report on the study is due sometime this summer.

In 1997 alone, says Orum, U.S. chemical facilities reported 38,305 chemical accidents to the EPA—roughly one chemical fire, spill, or explosion every fifteen minutes, he adds. Of those reported, more than 1,000 resulted in death or injury. Each year, 250 people die in chemical accidents, he says.

Many citizens depend on right-to-know laws. "Without community right-to-know provisions, a community is going to be severely lacking in information that has a direct impact on them," says Wilma Subra, a recipient of a MacArthur genius award for her work in helping citizens in Louisiana and other states clean up their toxic waste sites. "If you live near a facility, believe me, you want to know where those threats are."

7

The Private Sector Is Key in Cleaning Up Toxic Waste

Steve Lerner

Steve Lerner writes on environmental issues and is also the author of Beyond the Earth Summit: Conversations with Advocates of Sustainable Development.

One of the central debates over toxic wastes concerns public funding versus private initiative. On one side, advocates of public solutions want government oversight, regulation, and strict public accounting of the nation's hazardous wastes. Advocates of the private sector see individual initiative, motivated by idealism as well as profit, as the more effective solution to toxic cleanup.

The experience of William M. Haney and Christopher J. Nagel, of Molten Metal Technology, Inc. (MMT), lends strong support for the advocates of the private sector. Together, the two have developed a process that completely recycles hazardous waste without generating the hazardous byproducts of simple burning. Although the process is still controversial, MMT has attracted attention from environmentalists as well as investors looking for a moneymaking opportunity in the emerging "environmental industry." MMT also demonstrated a possible new paradigm for the hazardous waste industry: technologies that completely recycle industrial chemicals and byproducts and eliminate "toxic waste" from the economic equation altogether.

In a renovated helicopter factory located in an industrial park outside of Fall River, Massachusetts, Molten Metal Technology, Inc. (MMT), is developing a new way of recycling hazardous waste. If the new technology pans out, it will put this small town on the map for something other than being the place where Lizzie Borden allegedly gave her parents 81 whacks with an ax.

In the center of a huge, brightly lit room at the Fall River research facility stands a giant machine called a catalytic extraction processing (CEP) system. This Rube Goldbergesque contraption of metal boxes and pipes is advertised as being capable of digesting hazardous wastes and spitting out industrial-grade materials.

At its core the CEP is a closed-system cauldron of molten metal, usually iron, which is heated to some 3000° F. Any hazardous waste injected into the bottom of this molten metal bath is instantly dissolved. The catalytic and solvent properties of the molten metal break the molecular bonds that hold the hazardous chemical compounds together and reduce them to their elemental components. For example, polyvinyl chloride (PVC) wastes can be fed into the CEP and hydrochloric acid (HCl) and synthesis gases, or syngases (hydrogen and carbon monoxide) are recovered.

Controlling the CEP are technicians wearing blue uniforms, yellow hardhats, and safety glasses who manipulate the dissociated chemical compounds into new forms by adding materials to create desired products. The recovered materials are then partitioned into a gas phase, metal phase, or ceramic (slag) phase.

After injecting a waste into the molten metal, a portion of it is converted into a gas such as carbon monoxide or hydrogen, which can be filtered, separated out, and sold. Another part of the waste is extracted as any one of a number of metals such as chromium, cobalt, or nickel. The metal is concentrated into an alloy that can be sold or reused as feedstock for an industrial process. The remainder of the material, now rendered relatively benign, is partitioned into a ceramic, some forms of which can be sold as abrasives or aggregate [combining material].

Depending on how one defines the sector, the size of the environmental industry in the United States is anywhere from $100 billion a year to $170 billion.

Ancient alchemists would have appreciated this modern device for the way it transforms dross into valuable materials. Judged by today's standards, however, a new technology that processes hazardous waste is greeted with considerable skepticism by a public wary of technologies that typically spawn a new generation of unwanted byproducts. As a result, MMT faces an uphill struggle for acceptance. As one of a number of new environmental technology companies, it must not only prove it can recycle hazardous chemicals but also navigate a maze of environmental regulations. . . .

A controversial field

Molten Metal Technology is possibly the most widely known new company in the controversial field of hazardous waste management. Informed opinion about MMT runs the gamut: there are those who think it could revolutionize hazardous waste treatment, those who think it is a sham, and those who simply consider it a promising new technology with some advantages and some problems. But whether MMT lives up to its promise or fails, its story of entrepreneurship on the frontiers of environmental science is becoming increasingly common, as inventors and businessmen rush to get a piece of the market for green goods and services.

Depending on how one defines the sector, the size of the environmental industry in the United States is anywhere from $100 billion a year to $170 billion. One estimate places the number of jobs in U.S. environmental protection in 1992 at 4 million, or about 3 percent of total employment. Moreover, the international market for environmental goods and services could well reach $4 trillion for this decade, says Michael Silverstein, editor of *The Environmental Industry Yearbook*. And while most observers agree that the United States has recently lost its lead in the green market to Germany and Japan, Silverstein believes that this country, with its 30 years of experience in environmental cleanup, is still best positioned to develop the technologies that will ultimately dominate.

Starting young

William M. Haney III, president and CEO of MMT, is a young entrepreneur with substantial experience in tapping the power of green tech. Haney started young, incorporating his first environmental technology company, Fuel Tech Inc., in 1981 while only an eighteen-year-old freshman at Harvard. The company produces front-end air pollution control equipment designed to reduce nitrous oxides (NO_x) emissions. The technique, known as NOxOUT, which reduces NO_x emissions by up to 80 percent in the post-combustion phase, mixes dried urea [a chemical compound of carbon, nitrogen, oxygen, and hydrogen] with water and injects it into a flue gas of gas turbines, stationary diesel engines, power plants, and incinerators. "All of a sudden the pollution goes away," notes Haney, who sold his interest in Fuel Tech in 1987. It is now a company that has 75 plants operating and 200 coming on-line, he adds. Not bad for a freshman effort.

His second company, Energy BioSystems Corporation (EBC), founded in 1989, engineered a microorganism that extracts sulfur from crude oil and high-sulfur coal, permitting these fuels to be burned without generating sulfur dioxide. The microbe transforms the sulfur into sulfate, an inorganic salt which can be used as a fertilizer. EBC is currently estimated to be worth $90 million.

At this point Haney could have retired comfortably to a life of luxury, but this prospect did not appeal to him. "Walt Disney used to say 'I make money so I can make movies; I don't make movies so I can make money,'" he observes. "I want to use my money to deploy science to solve environmental problems." With this in mind Haney started another environmental technology company. Having attained the ripe old age of thirty-three, he launched his third and most ambitious venture, MMT, which some think has the potential to revolutionize the $12.2 billion U.S. waste industry.

MMT is on the cusp of a scientific revolution, Haney insists. He likens the position of environmental technologies today to the stage the biotechnology industry was at in 1980 just before it took off. "If I told you I was going to analyze a cow, clone the genes in her milk, add these cloned genes to an insulin shot, and make people who are dwarves full-sized, you would probably say, 'Haney, that sounds like a bit of a stretch,'" he says with a laugh. But someone did it and there is no reason why comparable wonders cannot be achieved in the environmental field, he adds.

Practical inventing

Haney's counterpart is the inventor of the molten metal process, thirty-eight-year-old Christopher J. Nagel, now executive vice president for Science and Technology at MMT, who sits down the hall from Haney in an office with a photo of a wild-haired Einstein staring down from the wall. Nagel's story is a classic tale of a young inventor who finds a practical application for what at first appears to be an obscure observation in an esoteric field of study.

His story begins in 1982 when, after graduating from Michigan Tech, Nagel landed his first job researching new energy conservation measures at the behemoth U.S. Steel works in Gary, Indiana. From his office—wedged between blast furnaces and [self-propelled] torpedo cars carrying hundreds of tons of hot metal—Nagel watched the plant produce hundreds of pots of molten metal slag every day, which were simply allowed to cool. He soon came up with the idea of using the heat from the molten metal to drive catalytic reactions that would release more energy, including a reaction that used old tires in reducing iron oxide to iron.

The combination of public pressure, strict standards, and the potential for liability can persuade a company to reduce or detoxify its hazardous waste rather than dump it.

In 1986 Nagel, who is the kind of guy who carries a card with a copy of the periodic chart [a table of chemical elements] in his wallet, left U.S. Steel (now USX) to do graduate work in chemical engineering at the Massachusetts Institute of Technology. There he developed the concept of using molten metal to drive reactions that would recover useful materials from hazardous waste. He described the technique, which he dubbed "catalytic extraction processing" (CEP), to the head of the licensing and patent office at MIT, John T. Preston, who introduced him to Haney. From the meeting of the inventor and the entrepreneur, MMT was born.

The non-dumping business

Hazardous waste treatment is a briskly moving sector among green-tech startup industries, and for good reason. According to the most recent figures from the Environmental Protection Agency's toxic release inventory, there were 37.3 billion pounds of hazardous wastes generated in the United States in 1992 (and the inventory only covers production-related wastes). Some 3.4 billion pounds of this toxic material is either released to the environment or sent off-site for disposal; the rest is managed in some way through energy recovery, recycling, or treatment. What makes this sector attractive to entrepreneurs is not so much the volume of waste, but the rules governing its disposal and the public clamor for cleanup. The combination of public pressure, strict standards, and the potential for liability can persuade a company to reduce or detoxify its hazardous waste rather than dump it. . . .

Among these new technologies for processing hazardous wastes, MMT appears to have attracted ample investments. Investors clearly think the CEP has great potential and Wall Street initially reacted enthusiastically to the prospects for this new technology. When MMT went public, Haney found little difficulty raising $105 million on the Nasdaq [stock] exchange. And MMT, which employs more than 300 people, has in hand orders for CEP systems from a number of major corporations. (Rather than processing waste commercially at its research facility, MMT aims to set up CEP systems on site at other companies. It may either own or operate the system itself, under contract for the client's waste, or sell the system to the client.) Clients include Hoechst Celanese, a Fortune 100 chemical corporation; Scientific Ecology Group, a subsidiary of Westinghouse Electric Corporation, the world's largest processor of low-level radioactive wastes; and M4 Environmental, a partnership MMT formed with Lockheed Martin.

MMT is also one of the most environmentally ambitious of the new hazardous waste processing companies. Its officials claim that the CEP can recycle 90 percent or more of a number of hazardous waste streams the company has placed a priority on, including PVC [polyvinyl chloride] wastes, electronic component wastes, and chlorinated solvents. They also claim that, in seven waste streams that are priorities for MMT, the CEP's destruction removal efficiency (DRE) is 99.9999 percent or greater. Haney says that the CEP, when hooked to the tail end of certain industrial processes, will create a closed-loop system, in which virtually nothing escapes and everything is reused.

The molten metal advantage

Heat-based treatment of hazardous waste is far from new. Incineration also uses a chemical reaction—oxidation during burning—to transform toxic compounds into more innocuous substances. But much of the trouble with incineration stems from the fact that fire does not deliver a constant temperature. By contrast, the temperature of molten metal can be held more constant, and molecules in a molten metal bath have more intimate, more complete contact with the metal than they would with a flame, says Dr. William Moomaw, professor of international environmental policy at the Fletcher School at Tufts University. As a result, he says, toxic compounds are wholly broken down in MMT's process.

"People still ask me if this new technology for recycling hazardous waste is too good to be true," says Haney, but the question is asked of him less often now that he has a track record for developing a string of technologies that help solve environmental problems. Sitting in his modern corporate offices in Waltham, Massachusetts, dressed in chinos and a denim shirt buttoned at the throat, Haney is brimming with confidence. "If in 1950 I told you that on a piece of silica the size of your thumbnail someone would produce something that would have the thinking capacity of 400,000 people for 400,000 days, what would you have said?" he asks with a smile. "If I had told you that I could develop a missile that would fly out of a silo in Nebraska and come down within a hundred yards of where I wanted it to land half a world away, would you have believed me? Yet we now accept that as perfectly normal. I would argue that this took a much larger leap of technological fancy than what we are doing at Molten Metal."

But there is no definitive answer yet on just how closely results match theory at MMT, and independent environmental specialists are far from seeing eye to eye on it.

Moomaw is one of the more sanguine. "The CEP system really is a breakthrough technology. It is one of the few new and original ideas to come along. As a chemist, the notion of taking complex molecules that happen to be toxic and breaking them down either into their elemental forms or into simple molecules like carbon monoxide, which can be burned as a fuel, is very attractive," he observes. Silverstein, author of *The Environmental Economic Revolution: How Business Will Thrive and the Earth Survive in Years to Come* (New York: St. Martins Press, 1993), is just as bullish. He sees MMT as part of a new field of industrial ecology, which is redesigning industrial systems along the lines of natural systems where there is no such thing as waste because every waste product becomes the feedstock for a different process. "When the leaves fall off the trees in autumn you do not see squirrels picking them up and stuffing them in Hefty bags and hauling them to a landfill. The leaves become a feedstock for vegetation that grows on the forest floor," he says. "What you have with MMT is the kind of process you see with more mature ecologies where there is no such thing as waste or pollution."

"What you have with MMT is the kind of process you see with more mature ecologies where there is no such thing as waste or pollution."

Marco Kaltofen, president of Boston Data Chemical Corporation, a firm which does envirommental investigations for law firms, trade unions, and nonprofit organizations, also sees MMT's process as having distinct advantages over incineration of hazardous waste. "First, MMT does not necessarily generate the huge volumes of gas that you get with incineration" he notes. With incineration, the higher the temperatures achieved to insure destruction of toxic chemicals, the greater the generation of gases that must be processed and cleaned. By contrast, the CEP process collects metals as solids instead of oxidizing them into vaporous fumes or gases, which then have to be scrubbed out in the [smoke-] stack.

This is a considerable advantage, says Kaltofen, who was previously director of the Citizen's Environmental Laboratory, a project of the National Toxics Campaign. But it does not make MMT a panacea for dealing with hazardous-waste. It just means that some subset of wastes will be amenable to this form of treatment, particularly homogeneous streams of industrial waste.

MMT's system is superior to incineration because you can run it cleanly and you have the opportunity to recycle hazardous wastes, Kaltofen continues. "But this is a process you can screw up," he cautions: "You have to know what is in the waste stream in order to create conditions in the CEP that will allow you to avoid the creation of unwanted byproducts." And therefore the success of the CEP system will depend on the company that buys it. "As long as the people running the CEP are responsible" and have a policy of keeping emissions as limited as possible, "then they will do well with this technology," Kaltofen says.

"I think Molten Metal has to be a step forward over incineration and landfilling," agrees Paul Connett, professor of chemistry at Saint Lawrence

University and editor of *Waste Not*. "It is a more rational approach than incineration where you put things through a high temperature, propel them into the air, and then do your best to catch as many of the propellants as you can. Incineration is a very unsophisticated chemistry; this sounds a bit more sophisticated." However, even Connett has reservations. "The only trouble of course is that this new technology may produce the mindset that you can continue to make toxic chemicals instead of placing the attention on front-end reduction of their use," he warns. . . .

Skepticism

Knowledgeable observers of the CEP remain divided on whether or not they think this new technology will prove widely applicable. Pat Costner, senior scientist with the toxics campaign for Greenpeace, is surveying innovative technologies for processing hazardous wastes. Where MMT is concerned she is on the side of the skeptics. "I personally think that every omnivorous [all-consuming] technology for treating hazardous wastes, such as MMT, has a common array of disadvantages," Costner says, adding that none of these innovative technologies will work on all waste streams. "This technology has been configured specifically for circumventing Resource Conservation and Recovery Act (RCRA) regulations," Costner asserts. She notes that for years the incineration industry claimed that nothing came out of its stacks except water, steam, and carbon monoxide. "How many years did we listen to that?" she asks. "Is this the same kind of ballgame?"

But Marco Kaltofen argues that while MMT should not be regulated as a recycling operation, it is nevertheless a valuable new technology. Fundamentally, he says, the business of this company is "making your waste problem go away without creating any environmental problems—and that alone makes it a technology worth encouraging."

"What MMT's story shows is how sophisticated the public has become environmentally," adds NRDC [Natural Resources Defense Council, a nonprofit environmental group] senior scientist Linda Greer. Any new technology will have to open up its data and undergo serious scrutiny, because "people now realize how important it is to get this stuff right." And Greer believes that "in a way that is the most hopeful sign of all. Whether or not it's MMT that helps us get there, public pressure is moving toward a sustainable society. Innovative technology has a big role to play in that transformation."

8

The Courts Should Decide Toxic Waste Disputes

Roger Meiners and Bruce Yandle

Roger Meiners is a professor of law and economics at the University of Texas at Arlington, and Bruce Yandle is the Alumni Professor of Economics and Legal Studies at Clemson University.

The federal record for protecting the environment is weak. New regulations and new statutes handed down from federal agencies and from legislatures are usually passed out of political motives and influenced by special-interest lobbying organizations. In addition, lawmakers often respond to the crisis of the moment and pass legislation that is harmful or useless in the long run. The best solution to environmental problems is the common law, the body of decisions and precedents handed down by the civil court system. In deciding on a case-by-case basis, judges and juries are much more likely to reach sensible solutions to the problems posed by air, land, and water pollution.

If the Republican-controlled Congress' efforts to pass assorted statutes that would affect Federal environmental controls succeed, there would be compensation for regulations that cause a substantial drop in the value of private property; cost-benefit and risk analyses may be required in devising new environmental regulations; and the Clean Water Act might be modified slightly. Environmentalists are claiming such reforms would gut environmental regulation and roll back 25 years of progress. Only Chicken Little would peer at the pending reforms and pronounce environmental collapse.

The move to protect private property rights was revolutionary 200 years ago, but hardly should be so today. As recently as during the Reagan Administration, the government paid when private land was desired to protect wildlife and wetlands. A requirement for benefit-cost and risk analyses was called for in *Reform of Risk Regulation: Achieving More Protection at Less Cost*, a study by the Center for Risk Analysis at the Harvard School of Public Health. This kind of reform is evolutionary, not revolutionary.

Reprinted from "Get the Government Out of Environmental Control," by Roger Meiners, *USA Today* magazine, May 1996. Copyright © 1996 by the Society for the Advancement of Education. Reprinted with permission from *USA Today* magazine.

Environmentalists know this. The plaintive cries heard from the environmental community reflect a deeper distress. Things environmental no longer are out of bounds for rational debate. When examined with emotions stripped away, the Federal record for protecting the environment is weak at best. While the air is cleaner than it was 25 years ago, more miles of rivers are swimmable, and lots of contaminated soil has been dug up at one site and moved to another, the gains have come at huge costs that could have been significantly lower. Enviromentalists know this too. Their enterprise is threatened, however, and that is why they are nervous. It is an interesting bit of history that, when the environmental statutes were written during the Nixon Administration, the environmental groups played little role. The biggest polluters spent years lobbying, so the environmental statutes were drafted in such a way as to not offend some of their interests. How times have changed.

As one review of government treatment of the environment summarized, "If we are serious about environmental quality, we should avoid relying heavily on government to provide it."

It is common knowledge that the government is a terrible steward of the environment. When a government controls all property, as it did in the former U.S.S.R., the result is that almost the entire country becomes a Superfund site by American standards. Yet, how much better is the U.S.'s record?

Recent General Accounting Office estimates are that the cost of cleaning up just the radioactive messes on Department of Energy land will run a minimum of $300,000,000,000 to as much as one trillion dollars. Government handling of such materials has been so careless that cleanup costs are nearly impossible to estimate. Private handlers of radioactive materials never allowed such massive problems to occur. The situation is not unique to the Department of Energy. The Army Corps [of Engineers], which claims to be the protector of wetlands, has been a major destructive force, as has the Forest Service, Bureau of Reclamation, Department of Agriculture, and other agencies that maintain they are environmental guardians. As one review of government treatment of the environment summarized, "If we are serious about environmental quality, we should avoid relying heavily on government to provide it."

The proposed revisions of environmental statutes hold the promise of reducing costs and targeting resources to higher-risk situations, all in the existing blueprint of command-and-control regulation. What if the slate were wiped clean and the Environmental Protection Agency (EPA) became a monitoring and research agency and returned environmental control to its pre-1970 status? Would we all drown in a sea of toxic wastes and suffocate from chemical-laden air? Hardly. In fact, we feel the environment is safer under the rules of common law (a part of state law) than under the control of one Federal agency.

Making the case for markets and the rule of law never has been easy. If people used to having their daily bread provided by government agencies are told to forget that and try the free market, would they cheerfully give up command-and-control? What if none of the people even could remember how markets under a rule of law worked? Discussing common law solutions to environmental protection encounters a similar challenge. In an uninformed debate, government delivery of environmental quality seems to win hands down.

Government planning is attractive because of its simplicity. Congress writes a law and the EPA enforces it with regulations about how much of this or that is legal or illegal. That is so much easier to understand than a free society relying on people to protect their rights through private arrangements of their devising that are enforced through the courts. The role of government is limited to providing a well-functioning judiciary that helps enforce contracts, protect property, and resolve disputes.

While those who support Federal command-and-control environmental regulation usually pay homage to the notion of private property, it routinely is asserted that governmental controls are needed to protect public health and the environment, because individual abuses of property leads to environmental destruction that injures other people and natural resources. That is, if not constrained by regulators, private property owners will impose pollution on others. Private control of land may be economically efficient, but the common law of property fails to protect the environment adequately.

Americans have become so used to centralized control, they cannot remember that, a generation ago the Federal government played just a small role in water quality. Instead, citizens and communities protected their waters, via enforcement of common law rights and through various state and local water quality regulations. The primary rights that were enforced were those, under tort law, not to suffer a nuisance or a trespass. Most states also enforced riparian law, which provides that all who have property abutting a waterway or body of water have the right to normal use of the water, but may not reduce its use and enjoyment by others and downstream users.

A few cases illustrate the process that was at work. In *Carmichael v. City of Texarkana* (1899), the Carmichaels owned a 45-acre farm in Texas, with a stream running through it, that bordered on the state of Arkansas. The city of Texarkana, Ark., built a sewage system connected to residences and businesses. The sewage collected in the city was deposited "immediately opposite plaintiffs' homestead, about eight feet from the state line, on the Arkansas side." The Carmichaels sued the city in Federal court in Arkansas.

The court found that the "cesspool is a great nuisance because it fouls, pollutes, corrupts, contaminates, and poisons the water of [the creek], depositing the foul and offensive matter . . . in the bed of said creek on plaintiffs' land and homestead continuously . . .," thereby "depriving them of the use and benefit of said creek running through their land and premises in a pure and natural state as it was before the creation of said cesspool. . . ." The Carmichaels were forced to connect their property to a water system to obtain water for "their family, dairy cattle, and other domestic animals, fowls, and fish." The cost of the water hookup and use was $700. They claimed the value of their property was reduced by $5,000, the lessened enjoyment of their homestead over the past two years was valued at $2,000, and the dread of disease was valued at $2,000. Besides the claim for damages, the Carmichaels also sued in equity for a permanent injunction against "said open sewer, cesspool, and nuisance."

The judge found that the city was operating properly under state law to build a sewer system, but that there was no excuse for fouling the water by the Carmichaels' property, regardless of how many city residences benefited from the sewer system. Citing other cases, the court found that the action at law for damages was proper, as was the request for an injunction:

Further, the court noted, "One who creates a nuisance through an inherently dangerous activity or use of an unreasonably dangerous product is absolutely liable for resulting damages. . . . "

"If a riparian proprietor has a right to enjoy a river so far unpolluted that fish can live in it and cattle drink of it and the town council of a neighboring borough, professing to act under statutory powers, pour their house drainage and the filth from watercloset into the river in such quantities that the water becomes corrupt and stinks, and fish will no longer live in it, nor cattle drink it, the court will grant an injunction to prevent the continued defilement of the stream, and to relieve the riparian proprietor from the necessity of bringing a series of actions for the daily annoyance. In deciding the right of a single proprietor to an injunction, the court cannot take into consideration the circumstance that a vast population will suffer by reason of its interference." The judge wrote: "I have failed to find a single well-considered case where the American courts have not granted relief under circumstances such as are alleged in this bill against the city. . . ."

In *Whalen v. Union Bag & Paper Co.* (1913), a new pulp mill polluted a creek. A downstream farmer, Whalen, sued the mill for making the water, which he used for livestock and crops, unfit for agricultural usage. The trial court awarded damages of $312 per year and granted an injunction against the mill to stop harmful pollution within a year. The appellate division denied the injunction and reduced the damages to $100. The court noted that the mill was an important economic asset to the area. It cost more than $1,000,000 and employed 500 people, which was worth far more than the water was to the plaintiff. The New York Court of Appeals reinstated the injunction, stating: "Although the damage to the plaintiff may be slight as compared with the defendant's expense of abating the condition, that is not a good reason for refusing an injunction. Neither courts of equity nor law can be guided by such a rule, for if followed to its logical conclusion it would deprive the poor litigant of his little property by giving it to those already rich."

To make clear that its decision went beyond a case involving serious destruction of water quality, the court cited an earlier Indiana court holding in a similar case, *Weston Paper Co. v. Pope*: "The fact that the appellant has expended a large sum of money in the construction of its plant, and that it conducts its business in a careful manner and without malice, can make no difference in its rights to the stream. Before locating the plant the owners were bound to know that every riparian proprietor is entitled to have the waters of the stream that washes his land come to it without obstruction, diversion, or corruption, subject only to the reasonable use of the water, by those similarly entitled, for such domestic purposes as are inseparable for and necessary for the free use of their land; they were bound also to know the character of their proposed business, and to take themselves at their own peril whether they should be able to conduct their business upon a stream . . . without injury to their neighbors; and the magnitude of their investment and their freedom from malice furnish no reason why they should escape the consequences of their own folly."

This holding does not mean that there could be no pollution. It meant that there was no excuse for uninvited pollution that caused real injury. To avoid water rights litigation, the mill owner could have contracted for riparian rights from downstream landowners or bought the land along the stream.

The primary problem caused by land pollution is water pollution due to seepage from improperly disposed toxic wastes. Hence, there are few cases that only concern the pollution of one's own land. After the Love Canal case in the late 1970s, Congress passed the Superfund law in 1980 to regulate the cleanup of toxic waste sites. How the common law might have dealt with this issue appears in a small number of cases that have occurred despite the Superfund law.

In *Village of Wilsonville v. SCA Services* (1981), the Illinois EPA, backed by the U.S. EPA, supported the right of a chemical waste landfill, which was alleged to be causing damage to a nearby village, to remain in operation. The Illinois Supreme Court found that the landfill caused groundwater contamination and that there could be a chemical explosion, given the disposal technique used. The court held the landfill was a public and a private nuisance; the village residents had been there first. The landfill was built with state and Federal approval, thus encouraging the landfill owner to ignore consequences to its neighbors. The court noted that toxic landfills were legitimate, but held that they had to be constructed so as not to impose costs on surrounding landowners who had not agreed to the intrusion: "Where individual rights are unreasonably interfered with, the public benefit from a particular facility will not outweigh the individual right, and the facility's use will be enjoined or curtailed." The court issued a permanent injunction against the landfill and ordered that the toxic wastes be dug up, moved, and the land restored.

In *New York v. Schenectady Chemicals, Inc.* (1983), the firm, during the 1950s and early 1960s, had hired an incompetent company to dispose of its toxic wastes. The hauler dumped 46,300 tons of chemical wastes in a swampy 13-acre site that drains into an aquifer serving thousands of persons. Prior to this action, it became clear that there was damage being done to some wells in the aquifer. Of the waste, 82.2% came from General Electric and Bendix (which already had agreed to pay up); Schenectady Chemicals (SC) contributed 17.8%. The state sued for SC to pay its share of the remediation costs, and asked for assorted penalties under New York's Environmental Conservation Law (ECL).

The court held that the ECL did not apply. The dumping had occurred before the permit system was established, so no penalties from that law were relevant. The cause of action that applies is the concept of public nuisance, defined by the Court of Appeals of New York as "A public, or as sometimes termed a common, nuisance is an offense against the State and is subject to abatement or prosecution on application of the proper governmental agency. . . ." Further, the court noted, "One who creates a nuisance through an inherently dangerous activity or use of an unreasonably dangerous product is absolutely liable for resulting damages, irregardless of fault, and despite adhering to the highest standard of care." SC was liable for its nuisance and was ordered to pay its pro-rata share of the cleanup expense.

In *Wood v. Picillo* (1982), neighbors sued a farmer who maintained a hazardous waste dump on his property. The plaintiffs claimed the dump emitted noxious fumes and polluted ground and surface waters. Holding for the plaintiffs, the Rhode Island Supreme Court overturned a 1934 decision that would have held for the defendant because groundwaters were "indefinite and obscure." The 1934 court had held that plaintiffs in pollution cases had to show that defendants should have "foreseen" the consequences of their action—that is, were negligent. The 1982 court held that, since 1934: "the science of groundwater hydrology as well as societal concern for environmental protection had developed dramatically. As a matter of scientific fact the course of subterranean waters are no longer obscure and mysterious. . . . We now hold that negligence is not a necessary element of a nuisance case involving contamination of a public or private waters by pollutants percolating through the soil and traveling underground routes."

Here, a rule of strict liability was imposed on polluters who cause damage to ground and surface waters. This standard of care is consistent with old common law tort rules imposing strict liability in case of hazardous materials. Liability is imposed if there is evidence of injury or potential to cause injury. Advances in knowledge of the effects of toxic substances means tougher standards today than in years past.

Few common law air pollution cases are found compared to water pollution. One reason is because most air pollution comes from multiple sources, making it more difficult to identify the defendant. The following cases, though, make it clear that the courts long have recognized liability for air pollution. Since passage of the Clean Air Act of 1970, the EPA has been the primary controller of air quality. That agency, and various state pollution agencies operating under EPA supervision, issue pollution permits to major pollution sources and determine pollution limits on vehicles and other emission sources.

In *Georgia v. Tennessee Copper Co.* (1907), the state, on behalf of its citizens, sued two companies that operated copper smelters in Tennessee near the Georgia border. Chief Justice Oliver Wendell Holmes noted that a public nuisance had been created because the "sulphurous fumes cause and threaten damage on so considerable a scale to the forests and vegetable life, if not to health, within [Georgia]. . . ." The plaintiffs argued that they recently had constructed new facilities which reduced the scope of the problem, but the Supreme Court held for Georgia. The Court held that the companies would be given a reasonable time to build more emission control equipment, but that, if such equipment did not reduce emissions adequately so as to protect plant life in Georgia, the state could ask for an injunction to shut the plants down.

The principled nature of the common law lies in its evolutionary and competitive nature. The weakness of statutes is in their being influenced by special interests and lack of competitiveness.

In 1915, the parties returned to the Supreme Court. Defendant companies showed that their new, very expensive equipment reduced emissions by more than half. Georgia argued that this was not enough and demanded that the plants be closed. The Chief Justice appointed a scientist from Vanderbilt University to spend six months, at company expense, studying the emissions and the likely effect of new controls. In the meantime, the Court ordered the companies to cut back production so as to reduce emissions further. Based on the evidence presented by the scientist, the companies either would be allowed to continue operation with more emission control equipment in place or, if that could not reduce emissions sufficiently, an order to shut the plants would be issued.

In *H.L. Renken et al. v. Harvey Aluminum, Inc.* (1963), landowners near an aluminum plant in The Dalles, Ore., sued on the basis of trespass and nuisance for air pollution emitted by the plant (fluorine and other toxic gases). Harvey Aluminum was the largest employer in the town, with 550 workers. During the 1950s, the plant had installed various emission control devices, including scrubbing towers and sprayers, which stopped over 90% of the emissions. Nevertheless, about 1,300 pounds of fluoride escaped during daily operations. The company asserted that nothing more could be done. Experts testified that emission controls could be tightened, but only at substantial cost.

The court ordered the company to pay surrounding orchards for damages to their crops caused by the trespass of the damaging gases. The company also was told to install the new emission control equipment within one year, or an injunction against any further emissions (*i.e.*, plant shutdown) would be issued, as requested by plaintiffs.

In *Bradley v. American Smelting and Refining Co.* (1985), the Bradleys lived on Vashon Island, Wash., four miles from a copper refinery run by American Smelting (ASARCO). They sued ASARCO, a New Jersey corporation, in Federal court in Washington for damages, charging trespass and nuisance from the deposit on their property of airborne particles of heavy metals from the company's smelter. The smelter had operated since 1905, was regulated by state and Federal air pollution laws, and was in compliance with all regulations. The gases that passed over and landed on the Bradleys' land could not be seen or smelled by humans—they required microscopic detection.

The Federal court, which would use Washington common law to determine the case, was uncertain as to what that law was since there were so few cases in the area. The court asked the Washington Supreme Court to tell it the status of that state's common law of nuisance and trespass as applied to air pollution.

The court held that ASARCO "had the intent requisite to commit intentional trespass." Even though no harm was intended, and even though ASARCO did not know the Bradleys, the company knew particles were being emitted from its facilities. Secondly, the court held that "An intentional deposit of microscopic particulates, undetectable by the human senses, gives rise to a cause of action for trespass as well as a claim of nuisance." Hence, the ASARCO emissions created a nuisance and a trespass. The court noted that, for a cause of action for nuisance or trespass to be successful, there must be "proof of actual and substantial damages." Under the statute of limitations, the plaintiffs had three years to file the action once the injury becomes known. It further held that the case was not prohibited by the Washington Clean Air Act, the state equivalent of the Federal Clean Air Act. Upon return to Federal court, the case was dismissed for lack of evidence of damage to plaintiffs or their property from the air pollution.

One cannot know how common law liability rules for pollution would have developed in the last several decades had their evolution not been largely precluded by statutes and regulatory controls. Nevertheless, other areas of law offer some clues. Recent advances in pollution control technology and in understanding the consequences of pollution, as well as changes in society's attitude about the acceptability of pollution, would have led to a rule of strict liability under the common law for polluters, as has occurred in product defect law. Even in its limited role, the common law often sets standards far tougher than those set by statutes.

The common law does not evolve the most economically efficient or even the most "just" rules. Still, the decisions of hundreds of independent judges, responding to thousands of cases filed independently by private parties seeking to protect their common law rights, are far more likely to produce sensible principles than are legislative bodies that generate rules greatly influenced by special interests, or rules that may reflect a crisis of the moment, but make little sense for the long term. The principled nature of the common law lies in its evolutionary and competitive nature. The weakness of statutes is in their being influenced by special interests and lack of competitiveness. Had citizens and states been allowed to pursue common law remedies, such as private and public nuisance actions against polluters, litigation would have resulted in a more considered approach of the consequences of alternative rules of law.

It is not clear that pollution regulations have produced better environmental protection than what might have emerged from the common law in the absence of Congressional interference to "solve" an alleged crisis. Of even greater significance is the loss of a key freedom—the control of private property—in favor of centralized control. It is not clear that the environment is better; it is clear that some freedom has been lost. The environment is protected better by individuals seeking to guard their rights than when such matters are determined by technologically driven command-and-control standards determined by legislators and regulators.

9

Properly Constructed Landfill Conversions Can Contain Toxic Waste

Jessica Snyder Sachs

Jessica Snyder Sachs is a contributor to Discover *and other magazines and writes on health and family issues.*

Modern urban areas generate immense amounts of municipal waste—much of it hazardous to human health. The latest trend in hazardous waste disposal is the landfill conversion, in which a city dump is landscaped and made over into a golf course or public park. New technologies allow landfill operators to compact and seal the underlying waste to prevent it from coming into contact with the air or water. At one such site, Live Oak on the outskirts of Atlanta, the landfill operator is considering using gas generated from the landfill to provide electricity. If done wrong, such projects can pose a direct threat to public health, but if done right, they can succeed in turning a longtime nuisance into a community asset.

During his 12 years at Englewood Golf Course in Colorado, superintendent David Lee has seen some goofy things pop out of the ground, such as wigs, bowling balls, and car bumpers. But pop-up junk here is something less than surprising: the course sits on a curvaceous mound of trash some 40 feet deep. In some places, all that separates the velvety green from the garbage is a few inches of sod.

Two years ago at a converted landfill called Renaissance Park in Charlotte, North Carolina, a soccer mom went after a stray ball that had fallen into an eroded hole around a light pole. To see in the shadows, she pulled out a pocket lighter. An exploding fireball blew her several feet back from the methane-filled hole. Fortunately, she suffered little more than minor burns and a bad case of the shakes. Signs discouraging open flames and smoking in all five of Charlotte's landfill parks were quickly posted.

There are far more tangible signs of the waste that lies just inches below the Renaissance landfill cover. On an afternoon after a gentle rain, the ground at the park's 18-hole golf course crackles like the sound of Rice Krispies. The noise comes from large patches of mud bubbling with gas. "It looks like polenta boiling on the stove," observes retired course superintendent Robert Orazi. But it smells like rotten eggs. Last year, Orazi gave up after six years of coaxing the grass and trees to grow on two feet of soil baked dry from the heat of rotting garbage below.

The course is also plagued by uneven settlement that dimples the fairways, tilts putting greens, breaks irrigation pipes, and turns cart paths into rolling "whoop-de-doos" only a dirt biker would love. Then there's the Blob, a foot-tall lump of wiggly amber-colored ooze creeping out of the fourth fairway. "We tried shoveling it; we tried covering it. It just comes back," says Orazi. Tests show "it" to be a kind of alga that feeds on the iron-rich liquid that seeps up from below. And pop-up waste? Among the scariest finds, says Orazi, are blood bags and syringes. More typical are the tires and rubber hoses that literally float up through the soil.

Sudden hazards

The hazards don't end with belches of garbage and gas. The heat of decaying trash can itself ignite the gases a landfill releases. That may have been the case when a six-foot flame shot from a crack near Renaissance's sixth green in 1989. Workers quickly doused it. But such landfill fires can spread underground for miles.

Several years ago, in Mountain View, California, an open-air amphitheater built over a landfill erupted in smoke during a Grateful Dead concert. The landfill was equipped with a gas extraction system, but the city had turned over the system's maintenance to the production company that ran the concert. When the production crew saw the smoke coming out of a crack in the ground, they cranked up the suction. The smoke disappeared, but the suction drew the fire underground and fueled it. Luckily, engineers arrived before anyone was hurt.

Closed structures, of course, are particularly susceptible to landfill gas. Without proper sealing and venting, methane can seep inside a building on or near a landfill and rise to explosive levels. That's what happened two years ago in a snack bar under construction on a landfill driving range in North Hempstead, New York. One night the water heater kicked on, igniting a fireball that knocked down the walls.

Despite the scare stories, over the past 20 years hundreds of municipalities and landfill operators have fashioned closed landfills into golf courses, parks, ball fields, playgrounds, even ski slopes. There is no national tally—largely because dumps, especially closed dumps, are considered local domain. And there is little regulation. "You don't need an EPA permit to play ball on a landfill," says Allen Geswein, of the Environmental Protection Agency's office of solid waste. "And given the current political climate, I wouldn't expect any moves in that direction."

Yet the need for more and bigger dumps won't go away. The United States generates some 209 million tons of municipal waste each year, over four pounds per person per day. Although no one knows exactly how many landfills reach capacity each year, the number is probably well over a hundred, and these monuments to waste cost money to maintain. Since 1993, for example, EPA regulations have required landfill operators to prevent their sites from leaking gas or polluted water for at least 30 years after they're closed (by then, according to theory, most of the gases from the decomposing garbage will have been released). The associated maintenance costs can reach hundreds of thousands of dollars an acre, which makes conversion to a revenue-generating facility like a golf course attractive—but problematic.

In 1993 the EPA also set some minimum standards for the design and operation of new landfills. Though aimed at reducing off-site pollution, these rules have the side effect of improving safety and stability on top of landfills as well. They require operators to screen waste for obvious chemical hazards and to refuse medical or toxic waste. Bulk liquids—such as sewage—are acceptable only if they have been solidified with soil or other stabilizers. Operators must also cover each day's garbage with a six-inch layer of dirt, which reduces the blowing away of trash and odors. The landfill's final cap, in turn, must consist of at least two feet of compacted soil.

Scary stuff

But only last year did the EPA make a move to control some of the gases that bubble to the surface of closed landfills. These gases are produced by the microbial food chain in the anaerobic, or oxygenless, environs of a landfill. Some bacteria, for example, degrade cellulose into sugar. Others eat the sugar, producing the acid that feeds gas-releasing bacteria. The result of their feast is a mix of methane (50 percent), carbon dioxide (40 percent), and nitrogen (9 percent), plus the trace contaminants that produce the foul smell of decay. None of these gases are particularly hazardous when allowed to dissipate in open air, says Martha Smith of the EPA's office of air quality planning and standards. It's the remaining 1 percent that includes some scary stuff.

Despite the scare stories, over the past 20 years hundreds of municipalities and landfill operators have fashioned closed landfills into golf courses, parks, ball fields, playgrounds, even ski slopes.

When bacteria degrade household cleaning products, solvents, paints, and pesticides, they generate vapors that include such nasty carcinogens as benzene, toluene, vinyl chloride, and a half dozen others. Vinyl chloride is a particularly toxic and persistent gas—persistent because it kills the very microbes able to dechlorinate and so detoxify it. The EPA's new rules require landfill owners to monitor and control these dangerous vapors in the air just above the landfill cover, keeping them within a safety margin of 500 parts per million. Control measures usually include an underground venting system that sucks toxic vapors and other landfill gases aboveground and burns them off.

Unfortunately, EPA regulations apply only to large landfills—typically those serving more than 100,000 households—that have been opened or modified since 1991. "This isn't to say that smaller and older landfills aren't of concern," says Smith. The EPA encourages individual states to set higher standards. California, for one, actually does, she adds. Moreover, whether from civic-mindedness or fear of liability, some of the nation's garbage giants are pioneering new designs for landfills and landfill parks that far exceed government standards.

The modern landfill

The 188-acre Live Oak Landfill and Recycling Center on Atlanta's outskirts is a far cry from the haphazard dumps of the past. Roughly the size of several football fields, it is one of the Southeast's largest land fills—handling some eight tons a minute, 4,500 tons a day. Opened in 1986 by Waste Management—which is the world's largest waste-disposal company, with some 140 landfills—Live Oak will reach capacity in 2001. After that it may begin a new life as a recreational facility with soccer fields and horseback riding trails.

Last December trash compactors at the Live Oak site were still spreading refuse on top of two of the three trash heaps that will end up 160 feet high. The first two mounds sit astride a central pit, where the operation's next phase will begin. Garbage will ultimately fill this pit, then start piling up and out like an inverted mountain against the sides of its sister peaks. The result will be a single flattened pyramid with a playable tabletop some five acres in size.

At the bottom of the still-empty central pit are seven layers of protective barriers for gathering and removing leachate—the polluted liquid from the decaying waste. The uppermost layer is a two-foot blanket of glistening white sand; not ordinary sand but grains manufactured to a specific size. If the grain sizes varied, they would pack together under the weight of the landfill, and smaller grains would fill in the holes between larger ones, preventing the runoff of leachate. Buried within this permeable, carefully milled sand is a horizontal pipe that will carry the leachate to a low-lying area. From there it will be pumped out of the landfill for disposal.

Directly beneath this layer of sand is a thin—.06 inch—sheet of high-density polyethylene (HDPE) plastic. Below the plastic lies a quarter-inch geosynthetic clay liner consisting of two fabric layers filled with a dry granular clay called bentonite. When wetted by, say, a leak in the overlying plastic sheet, the bentonite swells to form a tight, highly impermeable barrier.

The next layer down is the landfill's drainage system—a thick screen of heavy HDPE perforated pipes. Should any leachate reach this grid, it will drain to a low-lying pit. Leachate filling the pit will lift a float, which sounds an alarm signifying that the primary liner system has been breached. Live Oak operators can then draw the leachate out of the pit by applying suction. An added safeguard is a bottom layer of high-density polyethylene, which in turn lies on top of six inches of compacted clay.

Above these protective barriers, daily operations begin. Unlike the casually heaped dumps of the past, Live Oak conserves space by squeezing every last bit of air out of the garbage, creating a tightly compressed landfill structure. The garbage is sorted and distributed by size and compressibility, then ironed flat by 100,000-pound trash compactors that grind along on broad, cleated rollers. The compacting continues in two-foot layers until some 1,400 to 1,700 pounds of waste have been compressed into every cubic yard of space. Uncompacted, the same cubic yard would hold just 500 pounds.

At day's end, an eight-to-ten-foot stack of smashed waste is covered with dirt and crushed once more into a "cell." Imagining the landfill in cross section, the daily cells form continuous rows called lifts, which in turn become the landfill's horizontal tissues. Trash compactors grade the landfill's outer slopes to a 30 to 33 degree angle to maximize the structure's stability.

The continuous grading and compacting will greatly reduce the settling of garbage after the landfill is closed. More important, the compaction helps ensure that settlement is smooth and even. Though Live Oak Landfill may eventually settle by a dozen or so feet over the next 30 years, the overall shape and surface contours should remain roughly the same.

At five acres, Live Oak's upper surface is too small to be converted into a golf course, but had that been the plan, bulldozers have shaped the top layer of refuse into berms, curving fairways, and flattened greens. For the more modest plan of a ball field or equestrian center, the landfill's upper surface will be graded into a broad, gentle crown with just enough grade, about 5 degrees, to quickly slake off rain.

Before capping the landfill, Live Oak operators will install vertical pipes down through some 140 feet of trash to collect methane-rich gas. Other landfill operators have fashioned even more detailed gas collection systems, including a grid of flexible horizontal perforated pipes that snake through the trash, absorbing gas and feeding it to the vertical gas collection pipes.

Live Oak could generate .8 to 2.4 megawatts of power, enough continuous energy to serve perhaps 1,200 to 3,600 homes.

Although the EPA requires only that the gas vented from a landfill be flared, Waste Management is considering another plan for Live Oak. The gas might be drawn off to an on-site power plant and used to generate electricity. In this speculative scenario, the company estimates that for some five to ten years after closure, Live Oak could generate .8 to 2.4 megawatts of power, enough continuous energy to serve perhaps 1,200 to 3,600 homes.

The crowning touch, of course, will be the landfill's cap, the crucial barrier between its waste and park visitors of the future. At Live Oak, plans call for a composite cover combining natural and synthetic liners. The layer that lies directly above the waste will be an 18-inch layer of compacted clay. Workers will iron the clay with 60,000-pound drum rollers until it's virtually impermeable to water. Above this layer they will install a synthetic membrane like the plastic that lines the bottom of the landfill. High-density polyethylene is a popular landfill liner because it consists of strings of polyethylene molecules (CH_2-CH_2) thousands of carbon atoms long. The extreme length and stability of polyethylene's carbon backbone allows the molecules to pack tightly together like a crystal and so resist the assault of corrosive landfill leachate. However, this extreme density comes at the expense of flexibility. HDPE's brittleness is not an issue at the bottom of the landfill, where the membrane lies on top of solid ground. But the landfill cover must be able to flex as the garbage beneath it decays and shifts in its bed.

A little chemical manipulation provides the answer: add hexene (C_6H_{12}) to the polyethylene. Hexene's molecular structure prevents it from folding up into the neat, crystalline structure of the polyethylene, thus creating "lumpy," disorganized patches in the polyethylene matrix. This extra elbowroom between the tightly packed carbon chains produces a more flexible, less dense polyethylene. By adding pigments and stabilizers to the polyethylene, chemists can ensure that the membrane lasts upwards of 200 years.

To prevent water from pooling onto—and possibly breaking—the landfill cover, Live Oak engineers will install a drainage net just above the surface membrane. Rainwater seeping into this open grid will flow to the landfill's edge. The drainage net, in turn, will be covered with a synthetic textile, over which will be heaped two feet of soil, seeded with grass. The entire cover system, from compacted clay to topsoil, is designed to achieve an impermeability of a ten millionth of a cubic centimeter of water—a leakage rate of less than 147 gallons per acre a year.

When the landfill cover is finished, the top and bottom liners will be sealed together like a gigantic plastic bag. Postclosure maintenance, such as sealing up fractures or repairing leaks, will be costly. Although Waste Management is reluctant to confirm details concerning revenue, the cost of constructing and operating Live Oak—including buying the land and converting it into a recreation area—will reportedly total some $400,000 an acre. That's about $75 million, and it sounds staggering until one calculates revenues for the 188-acre landfill. With tipping fees of $32 to $35 a ton, that's as much as $157,000 a day.

The trouble with landfill parks

In reality, few landfill parks in this country are as well-financed and state-of-the-art as Live Oak. The typical scenario has been that of a cash-poor local government trying to convert an old, unregulated dump into landfill that can be used as a park. "All too often, county engineers simply dump dirt on the landfill, plant some grass, and say here's your recreation area," says Morton Barlaz, an environmental engineer at North Carolina State University. "Without a properly engineered cover and a methane collection system, you're going to have big problems."

Stories like these strike fear in the hearts of municipal attorneys. "The idea of putting people on a landfill makes me shudder," says Ann Moore, an assistant city attorney for Chula Vista, California. As a land-use expert, Moore has followed the landfill conversion trend for many years. "It was real fashionable a while back, and now a lot of cities are experiencing big problems," she says. Adds Barlaz, "There's always the risk that local governments won't have money for the high maintenance these parks demand. When budgets get cut, parks are the first to go."

Others argue that active use may simply be incompatible with the idea of keeping landfills sealed tight within a "dry tomb" of plastic. Bill

Sheehan, director of environmental biology for a landfill engineering company in Lawrenceville, Georgia, warns that even the most durable synthetic covers are likely to be punctured by plant or tree roots. The irrigation needed to keep parks green is another bugaboo. If the added water penetrates the landfill cover, it can overload leachate collection systems. This is a particular problem when irrigation pipes break under the strain of uneven settlement, as they often do.

Good examples

Still, with dumps filling and open space dwindling, landfill conversions are probably here to stay. And waste disposal companies can point to several thriving examples. Take Mount Trashmore Park in Virginia Beach. Created in 1973 from a 68-foot-high, 650,000-ton garbage heap, the park is now one of the area's most popular—especially with young children, who flock to the colossal wooden playground at its base. Another success is a 600-acre resort in Industry Hills, California, home of two championship golf courses. Methane from the underlying landfill is used to heat two Olympic-size pools and a hotel laundry in the adjacent Sheraton Conference Center. Then there's Riverview Highlands, a ski and golf resort built on a 600-acre garbage mound south of Detroit.

Some communities, in fact, have apparently overcome their reluctance and are ready to embrace their trash wholeheartedly. With nearly 6 million tons of refuse already in place, Virginia Beach is now drawing up plans for another landfill-based park to keep Mount Trashmore company—one more than twice as high and 18 times as voluminous as the original. After its makeover, the landfill will be dubbed City View Park, for an obvious reason—from its crest you will be able to see all there is to see. It's the biggest thing in town.

10

Some States Actively Court the Nuclear Waste Industry

Jim Woolf

Journalist Jim Woolf writes on environmental and local issues for the Salt Lake Tribune.

Lawmakers and the public are just beginning to grapple with the issue of radioactive waste storage. Toxic byproducts of nuclear power generation and weaponsmaking remain poisonous for centuries, in some cases millenia. A principal argument of nuclear opponents is the impossibility of assuring safe storage for such lengths of time, and the likelihood of an accident, a theft, or a natural disaster that would release radioactivity and possibly bring about a national catastrophe.

The problem hits home in South Carolina, where the Barnwell facility near the Georgia border has been accepting low-level radioactive waste for two decades. Although the facility has remained free of accidents, and generates tax revenues on the waste shipped in, many South Carolina environmentalists want to see it shut down. For the most part, local people have no problem with Barnwell and want the facility to remain open.

Khosrow Semnani, owner of Envirocare of Salt Lake City, is paying close attention to the debate in South Carolina. His company has made application to accept and store radioactive wastes at its disposal facility in Tooele County, Utah. Envirocare has competition: Private Fuel Storage and Safety-Kleen are also applying for radioactive waste storage permits in Tooele County. If these applications are accepted, Utah will become one of the country's principal dumping grounds for hazardous nuclear waste.

Governor Jim Hodges is tired of South Carolina being the nation's "dumping ground" for low-level radioactive wastes. He says it is time for someone else to assume that dubious role.

Reprinted from "Shopping Around for Nuclear Waste," by Jim Woolf, *The Salt Lake Tribune*, February 20, 2000. Reprinted with permission.

Like it or not, Utah is a strong candidate.

Hodges has asked the South Carolina Legislature to severely restrict the amount of radioactive waste going to the state's 235-acre Barnwell disposal site. It is generally believed some form of the bill will pass before lawmakers adjourn in June.

Assuming that happens, radioactive waste will begin piling up in warehouses around the nation. And Envirocare of Utah hopes to open its Tooele County disposal facility to that waste.

Envirocare, a private company owned by Salt Lake City businessman Khosrow Semnani, already accepts mildly radioactive waste at its landfill and has submitted licenses to expand its operation into the more dangerous material now going to South Carolina. The technical review of Envirocare's application is expected to last at least a year.

The company's proposal also requires the approval of the Utah Legislature and governor; neither has taken a position yet.

"I don't know very much about it," Governor Mike Leavitt conceded this past week. "This is the beginning of the process where we all are going to learn about it."

The top regulator of the Barnwell site urges caution. Virgil Autry, director of South Carolina's Division of Radioactive Waste Management, said Envirocare of Utah's proposal to move into radioactive waste is a big jump.

"You need to look at it very closely," he said. "You're going from stuff you can hold in your hand with no exposure to something that is very dangerous."

Environmentalists worry about the long-term risks of burying radioactive material in the moist, sandy soil of the South Carolina coastal plain.

Still, the waste at the Barnwell site has been managed safely, Autry notes, and Envirocare could develop the skills to do the same.

Envirocare's proposal comes at a time when Utah already is being considered for two other radioactive-waste facilities that would bring in waste from around the nation. The list:

—Private Fuel Storage, a consortium of eight public utilities, wants to build a temporary storage facility for highly radioactive spent fuel from nuclear power plants on the Skull Valley Band of the Goshute reservation in Tooele County. The waste from about half of America's power plants would be stored in above-ground casks until a permanent disposal facility is opened. Leavitt adamantly opposes this proposal, fearing once the waste arrives in Utah it never will leave.

—Safety-Kleen Corp. is seeking approval for a plan to dispose of mildly radioactive waste at its Grassy Mountain site in Tooele County. That facility now accepts only hazardous industrial waste.

"The Utah public needs to make a decision whether they want to become the nation's dump," said Preston Truman, an activist on nuclear issues and former member of the Utah Board of Radiation Control. "It is the responsibility of the state to make sure that discussion takes place in open and public meetings. We need an overall policy."

Hodges said the Barnwell facility has benefited America's nuclear industry and generated millions of dollars in profits for the education system. But two factors are prompting the first-term Democrat to pull back from this controversial business.

The first is fairness.

Congress in 1980 attempted to convince states to band together into "compacts" and build a series of regional disposal sites for low-level radioactive waste. But most states have refused, leaving only three sites accepting this material today. They are: Envirocare of Utah, which accepts only the most abundant but least radioactive types of waste; U.S. Ecology's facility at Hanford, Wash., which accepts the full range of low-level radioactive waste from the 11-state Northwestern Compact (including Utah); and the Barnwell facility, operated for South Carolina by Chem-Nuclear Systems, which accepts all low-level radioactive waste from the remaining states.

"For 30-plus years we have borne the lion's share of the responsibility" for disposal of America's radioactive waste, said Hodges. "Clearly South Carolina has done its share."

The other factor driving Hodges' initiative is that Barnwell is filling up.

A recent study finds that unless shipments of out-of-state waste are cut back, there soon will be no space for the waste from South Carolina's seven nuclear power plants. Only two other states generate a higher per capita share of their electricity from nuclear power than South Carolina.

"That would be the worst of both worlds," said Hodges. "Here we were assuming the responsibility to handle nuclear waste coming in from across the country, and then when our time of need came, there wouldn't be a facility open in the state for us to dispose of our low-level nuclear waste."

Hodges said safety and environmental issues are a concern to many South Carolina residents, but they were not among his top reasons for restricting the flow of waste. "I've seen no evidence over the years that it [the Barnwell facility] has been operated in an unsafe manner," he said.

Barnwell, located about 80 miles south of Columbia, has been at the center of an intense political debate in South Carolina for almost two decades. Environmentalists worry about the long-term risks of burying radioactive material in the moist, sandy soil of the South Carolina coastal plain, and they don't like having the dangerous waste transported along rural roads leading to the site.

The facility has a lot of supporters, too. Taxes on the waste shipped to South Carolina have generated almost a quarter of a billion dollars for state education programs in the past five years and most nuclear experts believe the operation poses minimal risks to workers, nearby residents or the environment. The disposal operation has near-unanimous support among Barnwell County's 21,000 residents.

"There are more important environmental issues in South Carolina, but Barnwell happens to be the state's highest-profile environmental issue," said John Clark, an aide to Hodges and a member of the executive committee of the South Carolina Chapter of the Sierra Club. "In the public's mind, if you want to do something good for the environment, you need to do something about Barnwell."

Such talk confirms the worst fears of David Cannon, a self-employed accountant who was snacking on chocolate chip cookies and milk at Anthony's, a crowded cafe in Barnwell. It is proof, he argued, that the governor is responding to political pressure from environmentalists to shut down the facility rather than looking at the facts and encouraging continued growth of a safe and profitable disposal industry.

"I get so frustrated," said Cannon. "Common sense and politics are like oil and water."

The Barnwell facility is located in gently rolling terrain dotted with small towns near the South Carolina-Georgia border. It is a pretty area with patches of pine forest and scattered fields of cotton and peanuts. The disposal site is several miles outside of Barnwell. It consists of a nondescript, two-story administration building with several smaller buildings used for laboratories, waste consolidation and truck cleaning. The waste is buried in a large, fenced area behind the buildings.

Residents of the area are familiar with radiation. Just down the road from Barnwell is an entrance to the U.S. Department of Energy's 354-square-mile Savannah River Site where five nuclear reactors produced materials used in construction of atomic bombs. The site now has 35 million barrels of radioactive wastes awaiting disposal.

Federal law divides radioactive waste into three broad categories and sets different rules for the disposal of each. One is low-level radioactive waste, which contains short-lived radioactive materials that will decay to harmless levels in 500 years or less. Most comes from nuclear power plants, but other sources include research laboratories and hospitals.

The other two main categories are high-level and transuranic waste that contain radioactive materials that will be dangerous for thousands or even millions of years. The main sources of these are spent fuel from nuclear power plants and the nuclear weapons program.

The U.S. Nuclear Regulatory Commission allows low-level radioactive waste to be disposed of through "shallow burial" in sophisticated landfills designed to isolate the material for five centuries. The high-level and transuranic wastes must be placed in mined, underground caverns where they are less likely to be disturbed by human activities over the millennia.

Low-level radioactive waste is subdivided into three categories. Class A waste contains radioactive materials that will decay to background levels in about 100 years. Most of the waste currently accepted by Envirocare of Utah is Class A. The next is Class B, which will decay in about 300 years. Class C waste takes 500 years to decay. Class B and C waste usually is more intensely radioactive than Class A.

Trucks loaded with Class A, B and C waste arrive daily at the Barnwell site. The least radioactive materials are transported in barrels loaded inside trailers. More dangerous waste is sealed inside specially built steel transportation casks attached to flatbed trucks. These casks shield workers and passing motorists from radiation and prevent the waste from spilling out in an accident. There have been several accidents involving vehicles hauling waste to Barnwell, but none has resulted in a spill.

Once inside the facility's gates, the waste is slipped into large concrete "vaults" that are buried in shallow trenches. The trenches are then covered with a specially engineered cap of clay, plastic, dirt and rock that is supposed to keep the enclosed waste dry for centuries. Size of the trenches and vaults varies depending on the type of waste. The more intensely radioactive waste is placed in narrow trenches where dirt walls shield workers while the waste is moved into place.

Envirocare of Utah proposes a similar vault system for the Tooele County site, said company president Charles Judd.

The only significant environmental problem at Barnwell has been associated with tritium, a radioactive isotope of hydrogen. Tritium moves easily in water and has been found escaping from some of the oldest disposal trenches and was found in storm runoff water last year, said David Ebenhack, vice president for community relations at Chem-Nuclear Systems.

Tritium and water are the biggest management challenges at Barnwell, said Autry, the state regulator. This is one reason disposal sites in arid regions such as Utah are more attractive than sites in wet areas. Barnwell receives an average of 50 inches of precipitation annually, while Envirocare's site receives about 9 inches to 10 inches.

Clark, the aide to Hodges, said Utahns can expect to see sophisticated political maneuvering as the Legislature and governor get involved in consideration of the Envirocare proposal.

In South Carolina, Chem-Nuclear is represented by the "best lobbyists money can buy, " he said. And they successfully fought several past attempts to close the Barnwell facility. The most recent was in 1995. But rather than close the site, they turned the situation around and convinced the Legislature to open it to the rest of the country by offering to pay a $235-per-cubic-foot tax on all waste coming into the state.

The tax, which was projected to generate as much as $140 million a year, was dedicated to the construction of schools and scholarships for needy students.

The $140 million figure was overly optimistic. The first year South Carolina received $110 million. The amount declined steadily as companies found new ways to compact their waste into smaller packages. Projected revenue in 2000 is around $43 million.

There have been several accidents involving vehicles hauling waste to Barnwell, but none has resulted in a spill.

"While it's not as much as we anticipated, it's real money and it has gone for the purposes outlined in the legislation," said Hodges.

Members of the Utah Legislature already have started informal discussions about how much Utah could squeeze from low-level radioactive waste going to Envirocare, said Judd. He thinks $235 per cubic foot is too high and would prompt many generators to simply store their wastes.

"If the rate is reasonable—which is something less than that [$235 per cubic foot]—volumes will stay up and the waste won't be all compacted down," said Judd.

While Envirocare is the most likely company to get the South Carolina waste, it is not the only contender. Waste Control Specialists has requested permission from the Texas Legislature to dispose of low-level radioactive wastes at its site near the New Mexico border.

However, the Texas Legislature does not meet again until January 2001. And if lawmakers approve, Waste Control Specialists still would need to submit a license application that would take another 6 to 18 months to process.

11

The High-Level Nuclear
Waste Program Is Misguided

James Flynn, Paul Slovic, Roger E. Kasperson, and Howard Kunreuther

James Flynn is senior research associate at Decision Research, a nonprofit, independent social science research institute in Eugene, Oregon. Paul Slovic is president of Decision Research. Roger E. Kasperson is professor of government and geography at Clark University in Worcester, Massachusetts. Howard Kunreuther is Cecilia Yen Koo professor of decision sciences and public policy and co-director of the Wharton Risk Management and Decision Processes Center at the University of Pennsylvania in Philadelphia.

For the past 40 years, nuclear waste has been accumulating at the nation's nuclear energy plants and nuclear weapons manufacturing sites. Soon the United States will have to make a crucial decision on how to deal with this waste—a decision that will affect nuclear policy for a long time to come. But the problem and the decision are being mishandled by the responsible government agencies, especially the Department of Energy (DOE, the agency given oversight over the nuclear industry). The DOE seems to be following a "decide, announce, and defend" approach to its decisions instead of considering and utilizing the research of scientists and the opinions and ideas of the public. The DOE's pet project, a national nuclear waste depository at Yucca Mountain, Nevada, should be recognized as a failure and abandoned.

The United States began producing high-level nuclear wastes (HLNW) more than 50 years ago in the course of developing its first nuclear weapons. Since then, military and civilian uses of reactors have generated a large quantity of such wastes, creating serious management and disposal problems. Even now, after more than 40 years of effort and the expenditure of vast sums of money, the United States does not have a technically sound or socially acceptable policy for dealing with its high-level nuclear wastes. Indeed, its current attempt to solve the problem by creating a national repository at Yucca Mountain, Nevada, is mired in conflict and destined for failure.

Reprinted from "Overcoming Tunnel Vision: Redirecting the U.S. High-Level Nuclear Waste Program," by James Flynn et al., *Environment*, April 1997. Reprinted with permission from the Helen Dwight Reid Educational Foundation. Published by Heldref Publications, 1319 Eighteenth St., NW, Washington, DC 20036-1802. Copyright © 1997.

71

This article takes a fresh look at the problem of disposing of high-level nuclear wastes. After recounting the lessons we have learned during the last four decades, it offers some specific recommendations for moving beyond the present impasse.

An unacceptable approach

A well-designed, effectively managed, and socially acceptable program for disposing of HLNW would be difficult under the best of circumstances. What makes the current situation so disturbing, however, is that the agency responsible for the program, the U.S. Department of Energy (DOE), is attempting to develop a repository at Yucca Mountain by unilaterally redefining the rules, standards, and procedures that were initially put in place to assure public safety and guide the program to success.

DOE's mandate is to study the Yucca Mountain site and determine if it can safely contain HLNW for thousands of years. What DOE wants to do, however, is build a repository—not just conduct a study. This was signaled earlier when DOE ordered a special machine to dig a 25-foot tunnel as part of a repository design rather than using existing equipment, which could provide up to an 18-foot tunnel and give perfectly satisfactory access to the underground area. The result was significant additional cost and delay, which added almost nothing to the purpose of what is still called the "exploratory studies facility" but which symbolically suggests engineering progress on developing a repository at Yucca Mountain.

More recently, DOE proposed new regulations for Yucca Mountain, which would eliminate the current factors for disqualification as well as doing away with socioeconomic, environmental, and transportation assessments until after waste storage has taken place.[1] In a letter to energy secretary Hazel O'Leary, Nevada governor Bob Miller complained that the new regulations were contrary to the requirements of the Nuclear Waste Policy Act, but apparently DOE expects Congress to pass legislation endorsing its changes or take on the difficult task of devising a new program with all the implications of delay and public controversy. The problem of meeting regulatory standards is not just going to disappear, however, and as the U.S. General Accounting Office has recently reported, there are a number of serious impediments to DOE's plans and those plans are unlikely to meet the needs of federal regulatory agencies.[2]

In another recent move, DOE abruptly announced a total revision of its transportation program under which it plans to contract with private companies to accept spent fuel and transport these wastes to a repository or interim storage facility. This would cancel more than a decade of work on transportation issues, which are of critical interest to 43 states, hundreds of local jurisdictions, and more than 50 million citizens residing near shipment routes. The Western Interstate Energy Board, an arm of the Western Governor's Association, called this privatization plan "an approach that was unacceptable in the 1980s, and it is unacceptable now." In a letter to DOE, the board said the new plan would "renege on commitments" and constitutes a "shirkiner of responsibilities to states and the public."[3]

The model that DOE is following, the so-called decide-announce-defend (DAD) approach, failed badly at Lyons, Kansas, 30 years ago. That failure, and the long decade of confusion, delay, and frustration that preceded passage of the Nuclear Waste Policy Act (NWPA) in 1982, show clearly that the DAD model does not work. In fact, that model was implicitly rejected with the enactment of NWPA, which provided for regulation by the Nuclear Regulatory Commission and the Environmental Protection Agency, along with oversight by affected states and Indian tribes, created an intricate architecture of provisions for achieving fairness (including a requirement that DOE "consult and cooperate"); and established a process for building public understanding and support. However, DOE has been completely unable to develop alternatives to the DAD approach except in its public relations statements.

[Oak Ridge National Laboratory director] Alvin Weinberg concluded that he paid "too little attention to the waste problem. Designing and building reactors, not nuclear waste, was what turned me on."

A new approach to managing and disposing of HLNW is urgently needed. The current program should be halted and its unrealistic deadlines abandoned. Claims that a crisis in safe storage of HLNW justifies DOE's rogue program should be thoroughly evaluated. There appears to be ample time to create a safe HLNW program and to proceed in a manner that wins public acceptance and support. Two developments over the past decade are encouraging in this regard. First, new technologies such as dry-cask storage will enable the wastes to be stored for a century or more at existing reactor sites or other temporary facilities. Second, attempts to site disposal facilities (including those for low-level and transuranic wastes), along with experience in siting other hazardous facilities, have improved understanding of what is necessary to win public acceptance. Attention to safety, environmental protection, trust and confidence in program managers, and adherence to high standards of public values such as fairness, equity, and accountable decision processes are essential.

A difficult problem

Before the passage of NWPA in 1982, the prevailing opinion of those managing nuclear power was that disposal of HLNW was not a difficult or serious problem. However, the task turned out to be far more challenging than the experts could even guess. In reflecting on his years as director of the Oak Ridge National Laboratory (ORNL), Alvin Weinberg concluded that he paid "too little attention to the waste problem. Designing and building reactors, not nuclear waste, was what turned me on." If he could do it over, he would "elevate waste disposal to the very top of ORNL's agenda."[4] Alas, some experts and technical managers still maintain the old perspective that HLNW is a "rather trivial technical problem."[5]

In response to NWPA, in 1983 DOE embarked on programs to develop a centralized Monitored Retrievable Storage (MRS) facility in Tennessee; to identify and "characterize" (study) three first-round sites in the western United States for a permanent repository; and to identify a set of potential second-round repository sites in the eastern part of the country. Recognizing that most nuclear power stations were in the East, NWPA called for geographic equity, with both a western and an eastern repository site. This ambitious effort failed in many ways, most dramatically in 1986 when DOE abruptly suspended work on an eastern repository site under pressure from congressional representatives in an election year.[6]

Lessons from NWPA's early years

A wide range of problems were revealed in the first five years after NWPA was passed. First, obtaining reasonable assurance that a repository would protect human health and the environment was much more complex and uncertain than anticipated, and the public had great concerns about all aspects of handling, transporting, and storing HLNW.[7] Second, public acceptance was not achieved and the program ran into vigorous opposition from citizens, communities, and states at each potential site.[8] Third, relations between DOE and states and communities were contentious and often bitterly adversarial, as shown by the dozens of legal actions filed against the department in federal court.[9] Fourth, management and administration of the program were chaotic and inadequate.[10] (Schedules slipped and essential program activities such as quality assurance failed to meet regulatory requirements.[11]) Fifth, program costs soared.[12] Sixth, as information was put together for the environmental reports intended to compare first-round candidate sites, it became increasingly clear that scientific, engineering, and management problems would be difficult and contentious.[13] After five years of effort, the program was clearly not working.[14]

The basic principles of NWPA included a strong emphasis on public health and the environment, with independent regulation and state oversight; fairness and equity in process and outcome; "consultation and cooperation" with states and affected Indian tribes; and scientific and technical work of high integrity to determine the best sites. Combined with the other failures, DOE's ineffectiveness in putting these principles into effect prompted Rep. Morris Udall (D-Ariz.), one of the lead authors of NWPA, to call for a program moratorium and a thorough review in 1987.

An easier alternative was to amend NWPA and reduce the repository program to something DOE could accomplish. Sen. J. Bennett Johnston (D-La.), generally credited as the author of the 1987 amendments act, avoided the sort of open policy review that had preceded passage of NWPA and confined the substantive discussions to committees favoring a quick fix at Yucca Mountain. The amendments were steered through a complex and often byzantine process and eventually passed as part of the Omnibus Budget Reconciliation Act of 1987.

The NWPA repository selection process, with all its provisions for technical integrity and its measures to assure fairness, was abandoned. The second-round site selection program and the Tennessee MRS projects were canceled. Study of first-round sites except for Yucca Mountain was halted. In and out of Congress, the amendments were recognized as an opportunistic choice based on Nevada's weak political position in

Congress. (Nevadans called these amendments the "Screw Nevada" bill.) Equally seriously, the amendments did not address the underlying reasons for the failure of the original program, perhaps because no real review was conducted to understand why the program had collapsed.[15]

Focus on Yucca Mountain

As a result of the 1987 amendments, a repository at Yucca Mountain became the only option for storing the nation's HLNW. If DOE could not construct this facility, Congress would have to develop a new program. Nothing in these amendments effectively addressed public opposition, the lack of trust and confidence in DOE, the scientific uncertainty surrounding the program, DOE's management problems and the possibilities of other management options, the difficulties presented by intergovernmental conflict, or the rationale and priority of the legislated waste acceptance schedule. The only response to these pressing problems in the amendments was the congressional mandate to look solely at Yucca Mountain.

DOE's mission, although still cloaked in the terms of . . . "reasonable assurance" of safety, became fixated on the transfer of HLNW from the reactor sites to Yucca Mountain.

The prospects for protecting human health and the environment did not improve with the exclusive focus on a single site. The optimistic hunch that Yucca Mountain would prove to be a suitable site, as urged by science writer Luther J. Carter in his influential 1987 book,[16] began to fade along with the idea that DOE could provide the assurances required by regulators and the public. Then, in 1990, the National Academy of Sciences' Board on Radioactive Waste Management (BRWM) reported that the existing program was "unlikely to succeed" and that there was a need to "rethink the program" as a whole.[17]

Despite some advantages of the federal program, BRWM found one overriding disadvantage: Science cannot "'prove' (in any absolute sense) that a repository will be 'safe' as defined by EPA and the USNRC regulations." The board identified two reasons for this: the "residual uncertainties" of a repository preclude DOE (or anyone else) from making the required assurances, and "safety is in part a social judgment, not just a technical one." The technical problem turned out to be anything but trivial. BRWM suggested an "alternative approach" that would be more flexible, incremental, and adaptive. While the "ultimate performance" goal could be specified, the alternative approach would enable program managers to learn as they went along, use expert opinion to address issues of uncertainty, and react to problems rather than trying to anticipate them.[18]

Meanwhile, DOE made little progress in gaining public trust and confidence and in asserting management control over the program.[19] The nuclear industry and Congress were greatly concerned by the combined lack of program progress and escalating costs. Site studies were estimated to cost $60 to $80 million in 1981; by 1987, the estimates had risen to $2 billion for each of three planned study sites by 1987 and to $6 billion for the single site at Yucca Mountain by 1994. Congressional responses to these figures forced DOE to reduce its efforts and obligations at every turn to maintain some semblance of adequacy in the Nuclear Waste Fund.

The failure to understand and address the problems revealed during the first five years of the HLNW program and the abrogation of the basic principles in NWPA meant that the program as it was revised by the 1987 amendments act was seriously flawed. DOE's mission, although still cloaked in the terms of site characterization studies and "reasonable assurance" of safety, became fixated on the transfer of HLNW from the reactor sites to Yucca Mountain. Since 1988, DOE has been struggling to site an HLNW storage facility, churning through a series of program revisions and strategies, seeking ever more lenient standards and looser regulations, resisting demands from nuclear power utilities for early removal of their HLNW, fighting states and local communities who are afraid of the transportation and storage risks, striving to reduce program costs that constantly threaten to outstrip future Nuclear Waste Fund revenues, and attempting to defer licensing accountability for the ultimate safety of a repository at Yucca Mountain into the distant future.[20] DOE's implicit strategy is evident: If it could avoid facing the regulatory, cost, and safety issues long enough, a repository or interim storage facility could be developed at Yucca Mountain as a *fait accompli.*[21]

DOE's evolving, opportunistic approach has presented the nation with two major problems. First, it undercuts important performance goals and leaves the nation with a waste management program that changes the rules of the game as it goes along. (The department's management of radioactive wastes at the nation's weapons complex, where it set its own environmental and safety standards, suggests the dangers of such a strategy.) Second, if acceptable safety or risk "is in part a social judgement," then a new social judgment is clearly called for. Widespread lack of public trust, confidence, and support for the existing program make it clear that the 1987 amendments do not entail a societal mandate.

Ultimate goals and social judgments

What are the basic terms of a societal judgement? The National Environmental Policy Act of 1969 and subsequent legislation, such as the Clean Air Act and the Safe Drinking Water Act, expanded the standards of health and environmental protection and allocated responsibility for meeting those standards to federal agencies and those in state and local governments. NWPA attempted to structure the HLNW program so that it would conform with the spirit of the nation's environmental laws (although it also made special provisions to expedite the licensing and permitting process). EPA determined that a repository should provide containment for at least 10,000 years and that DOE must provide assurances that this ultimate goal would be met. However, these provisions require precise knowledge of future societal and environmental conditions over very long periods of time, thereby posing questions of great complexity and considerable uncertainty.

Much of the research done by DOE has addressed the physical conditions at Yucca Mountain. Preliminary as they are, these studies reveal a number of complex problems. The major radiation pathways from an HLNW repository to the environment are air and water. The geologic structure at Yucca Mountain is so fractured and faulted that it would not meet EPA standards for the release of carbon-14, which as a gas could easily escape to the surface.[22] Congress responded to this disqualifying condition by mandating less stringent methods for calculating the radiation risks for a Yucca Mountain repository in an amendment to the Energy Policy Act of 1992. The recent discovery at repository depths of chlorine-36, an isotope produced by atmospheric nuclear bomb tests, indicates that water migration from the surface has occurred in less than 50 years, contrary to expectations. The existence of this water pathway could compromise protection of HLNW within a repository.[23] Daniel Dreyfus, the director of the Office of Civilian Radioactive Waste Management (OCRWM), admitted that "a strict interpretation of the regulations would mean a disqualifying condition 'unless you can change the regs.'"[24]

The southern Nevada area is a young, geologically active environment. The Ghost Dance Fault, which cuts directly through Yucca Mountain from south to north, is of particular concern.

Seismic and volcanic hazards pose further problems.[25] On 29 June 1992, an earthquake of magnitude 5.6 occurred at Skull Mountain, just 12 miles from the repository site, seriously damaging DOE's project buildings at Yucca Mountain. This was followed on 17 May 1993 by an earthquake of magnitude 6.0 south of Bishop, California, about 100 miles west of Yucca Mountain. The southern, Nevada area is a young, geologically active environment. The Ghost Dance Fault, which cuts directly through Yucca Mountain from south to north, is of particular concern. As government geologists explained in May 1993, they do not know when this fault was last active or whether it is in a zone connected to other faults.[26]

Yucca Mountain was created from volcanic ash compressed into the rock ("tuff") that is the geologic medium for the repository. The Crater Flat area, which includes one dormant volcano about seven miles from Yucca Mountain, is a source of potential volcanic activity.[27] According to Geomatrix, a DOE contractor, there are "uncertainties in models that describe the future locations of volcanic events and models that describe the temporal distribution or rate of events."[28] In early 1995, DOE assembled a panel of experts and elicited their opinions about how likely future volcanic eruptions are. The panel concluded (in a report that was released in June 1996) that "the probability of a volcano erupting through the repository during the next 10,000 years is about 1 in 10,000."[29] Eliciting opinions as substitutes for data and predictive models saves DOE money, time, and research effort, but how well this use of subjective judgment defines the risks and uncertainties at Yucca Mountain is an open question. According to a report in *Science*, DOE will attempt to address other scientific and engineering uncertainties at Yucca Mountain in the same manner,

calling on expert panels that have been put through an extensive training program and coached to reach a consensus.[30]

In the face of massive technical uncertainty, contractors working for
DOE at Yucca Mountain are put in an equivocal position. An example of
how one DOE contractor attempted to convert unresolved uncertainty
into a benefit is demonstrated by a site suitability study which stated (in
somewhat convoluted language) that uncertainty and lack of evidence
fail to show that the repository will not meet qualifying conditions.[31]
This curious logic was then endorsed by DOE's Yucca Mountain project
director as evidence that the repository program was on the right track.[32]

Responding to complexity and uncertainty and taking into account
the surprises that certainly will occur over a 10,000 year period requires a
new level of adaptability—that is, the ability to adjust to technological
and social change as well as to unforeseen circumstances. Important questions abound in this area. Should DOE, on its own, make the necessary
adjustments? Can DOE define what is acceptable to society, or is it too
committed to Yucca Mountain to conduct objective studies and achieve
real public acceptance?

Another legislative fix

Some people think the government is now too far along with the Yucca
Mountain project to back away. Given where we are, they say, lower standards should be adopted so that we can move ahead. Science writer
Luther J. Carter recently voiced this view, stating: "Today the nation simply does not have the political stomach for another search for sites, for either a geologic repository or a surface storage facility." Besides, Carter
says, even if Yucca Mountain has problems, other sites might not be any
better.[33] However well such casual observations might apply to ordinary
government programs, they seem out of place for a first-of-its-kind
HLNW repository that must ensure safe isolation of radioactive wastes for
10,000 years or more.

The problem for proponents of the Yucca Mountain project is that to
save it they will have to change the law and the rules. An attempt to "fix"
the Yucca Mountain program was passed by the Senate in 1996 (Senate
bill 1936) but died in the House under the threat of a presidential veto.
The terms of that bill—which has been described as a nuclear industry
wish-list—have now been incorporated into Senate bill 104, introduced
by Sen. Frank Murkowski (R-Alaska) on 21 January 1997. The primary
strategy is to pass a new law that weakens the standards again and attempts to force acceptance of both an interim storage facility and a repository at Yucca Mountain. Senate bill 104 provides rewards for small, local
jurisdictions in Nevada that are viewed as allies of DOE's program and
completely removes the state of Nevada as a participant.

Special standards for Yucca Mountain would assess radioactive doses
to the public for 1,000 rather than 10,000 years, even though HLNW will
remain hazardous well beyond 10,000 years.[34] An "average member of the
general population in the vicinity of Yucca Mountain" could receive
doses of radiation as high as 100 millirems (mrem) per year. The Cana-
dian standard, by contrast, is less than I mrem per year, while the (U.S.)
National Academy of Sciences' recommendation is 2 to 20 mrem per year,

similar to the standards in other countries. The New Mexico Waste Isolation Pilot Project (for disposal of less dangerous transuranic wastes) is limited to 15 mrem per year, and the NRC's low-level waste (the least dangerous radioactive wastes) standard is 25 mrem per year.[35]

Regulatory concern about human intrusion at a repository would be removed by having NRC *assume* that the engineered barriers and postclosure oversight will prevent human intrusion and exposure. John Cantlon, chairman of the Nuclear Waste Technical Review Board, clearly stated the problem confronting us here: "There is no scientific basis for predicting the probability of inadvertent human intrusion over the long times of interest for a Yucca Mountain repository." But rather than suggesting scientific study of this important problem, he concludes that "intrusion analyses should not be required and should not be used during licensing to determine the acceptability of the candidate repository."[36] Inadvertent and deliberate intrusion appear confounded here, but in any case it is curious to hear eminent nuclear experts argue that the current inability to assess a risk is a reason for eliminating it from consideration in the licensing process.

A new approach . . . is needed. The first step is to halt the current program and drop the unrealistic schedule.

It is not only scientific uncertainty and the possibility of human intrusion that present risks to human health and the environment at Yucca Mountain. Having DOE as the manager of the HLNW program entails risks that are seldom acknowledged but deeply troubling. Some nuclear industry supporters claim that storage of HLNW at nuclear power stations is an urgent problem. They demand that these wastes be transferred from places where they have been managed safely for 40 years or more, and where they can be securely stored for another century, to a federal facility. It is ironic that they wish to hand them over to DOE, which is now struggling to clean up—at a cost that may reach hundreds of billions of dollars—massive radioactive contamination at weapons sites across the United States.

Where do we go from here?

A new approach to managing and disposing of HLNW is needed. The first step is to halt the current program and drop the unrealistic schedule. Flexible and realistic timetables would allow time for adequate research on technical and social problems and comparison of alternative approaches, such as seabed disposal.

We should attend to the experiences of other countries, which have developed much more flexible schedules and research programs, learning and adapting when faced with public opposition to their HLNW plans. The programs in Canada, Sweden, the United Kingdom, and France are certainly not without problems, but the governments of those countries have also been willing to change—often dramatically—to address public concerns.[37]

The experience of these countries suggests quite strongly that the decide-announce-defend approach and the use of central government authority to override public concerns are not likely to be successful. France used such an approach prior to 1990, when widespread public opposition led to a temporary moratorium on the search for disposal sites. Similarly, the United Kingdom scrapped its DAD approach in favor of "The Way Forward," a more open program with involvement by a wide range of stakeholders. In Sweden, a low- and intermediate-level waste facility and an interim HLNW storage facility have been opened and a specially designed transport ship put into service. This was accomplished with a program that listened and responded to critics and host communities to achieve a broad public consensus. The United States clings to its rushed schedule and the aim of "getting the waste into the ground" with a DAD strategy that is now widely discredited across advanced industrial societies. Failure to learn from its problems and restructure its program, as other societies have done, remains one of the most puzzling attributes of the U.S. experience.

Dry-cask storage could be used at reactor sites or at centralized, perhaps regional, locations. There are some questions about such interim storage, but it provides necessary additional time for research and public consultation and it is considerably less costly than the pursuit of a troubled geologic repository program.

In looking toward permanent disposal, it will be important to examine more than one site and more than one option. Keeping the available options open until a firm social, political, and technical consensus develops is a prudent way to address the issues of uncertainty and intense public opposition. A new process should employ a voluntary site selection program. Congress should mandate that no community be forced to accept a repository against its will, and potential host communities should be encouraged and empowered to play a genuinely active role in the planning, design, and evaluation of any HLNW facility, especially a repository. A voluntary process should be structured to acknowledge the legitimacy of public concerns, provide for public participation, develop agreed-upon procedures, commit to openness and fairness (including mitigation and compensation where appropriate), and empower state and local communities in decisions about safety.

Experience with other noxious facilities has revealed that the siting process is most likely to be successful when trust is established between the developer and the host region. This trust is likely to emerge only if the public has an opportunity to participate fully in the siting process.[38] The credibility of the U.S. HLNW program must be established. This requires a new organization to replace DOE, one capable of fulfilling the management and oversight roles required to address this complex and uncertain program. A new and different approach is needed—one committed to implementing a restructured program in an open, consultative, and cooperative manner, and one that does not seek to deny or avoid the serious social, political, and economic problems.

Summary

The time is long overdue to recognize publicly that Yucca Mountain is a failed program. We can learn from this failure, however. We now know, for example, how important public acceptance and trust are to the success of any HLNW program. We have learned that intergovernmental cooperation must be earned and not simply mandated by Congress; that funding and budgeting will have to be reconceptualized and restructured; that the scientific and engineering challenges of locating, building, and operating a geologic repository are complex and will involve significant levels of uncertainty; and that the ultimate goal of protecting human health and the environment can easily be eroded unless continuous societal oversight takes place. The existing program has demonstrated that it is fundamentally incapable of addressing the problems of HLNW. The next essential step is to pause and explore the proper questions, learn from past mistakes and the experience of other countries, and begin the difficult task of creating a socially acceptable solution in the United States.

Notes

1. U.S. Department of Energy, "General Guidelines for the Recommendation of Sites for Nuclear Waste Repositories (Proposed Rule)," *Federal Register* 61, no. 242 (16 December 1996): 66157–69.

2. U.S. General Accounting Office, *Nuclear Waste: Impediments to Completing the Yucca Mountain Repository Project*, GAO/RCED-97-30 (Washington, D.C., 1997).

3. R. Moore and D. Nix, cochairs. Western Interstate Energy Board, letter to R. Milner. Office of Civilian Radioactive Waste Management, U.S. Department of Energy. 13 January 1997. The Western Governors Association includes twelve western states.

4. A. Weinberg, *The First Nuclear Era: The Life and Times of a Technological Fixer* (New York: American Institute of Physics, 1994), 183.

5. B. Cohen, *The Nuclear Energy Option: An Alternative for the 90s* (New York: Plenum Press, 1990), chapter 11. Also see L. J. Carter, *Nuclear Imperatives and Public Trust* (Washington, D.C.: Resources for the Future, 1987), pages 9–10, for a number of similar opinions expressed by nuclear experts during the period before and after congressional action on NWPA.

6. Carter, note 5 above, provides a good historical account of the development of the HLNW program and the difficulties of its first years of operations.

7. An overview is presented in J. Flynn et al., *One Hundred Centuries of Solitude* (Boulder, Colo.: Westview Press, 1995), chapter 4. Also see G. Jacob, *Site Unseen: The Politics of Siting a Nuclear Waste Repository* (Pittsburgh, Pa.: University of Pittsburgh Press, 1990); and National Research Council/National Academy of Sciences, Board on Radioactive Waste Management, *Rethinking High-Level Radioactive Waste Disposal* (Washington, D.C.: National Academy Press, 1990).

8. The scope of these responses is indicated in R. Dunlap, M. Kraft, and E. Rosa, eds., *Public Reactions to Nuclear Waste: Citizens' Views of Repository Siting* (Durham, N.C.: Duke University Press, 1993).

9. U.S. General Accounting Office, *Nuclear Waste: Quarterly Report on DOE's Nuclear Waste Program as of September 30, 1987*, GAO/RCED-88-56FS (Washington, D.C., 1988), appendix 1.

10. Nuclear Waste Strategy Coalition, *Redesigning the U.S. High-Level Nuclear Waste Disposal Program for Effective Management*, draft report (St. Paul, Minn.: Minnesota Department of Public Service, 1994). The coalition was made up of state regulators, utility executives, and attorney general representatives. This report reviewed 18 federal government documents on DOE management problems during the period 1982 to 1994 and recommended that DOE be removed from HLNW management and its duties assigned to a new federally chartered corporation.

11. U.S. General Accounting Office, *Nuclear Waste: Repository Work Should Not Proceed Until Quality Assurance Is Adequate*, GAO/RCED-88-159 (Washington, D.C., 1988).

12. U.S. General Accounting Office, *Nuclear Waste: Information on Cost Growth in Site Characterization Cost Estimates*, GAO/RCED-87-200FS (Washington, D.C., 1987).

13. See D. Easterling and H. Kunreuther, *The Dilemma of Siting a High-Level Nuclear Waste Repository* (Boston. Mass.: Kluwer Academic Publishers, 1995), chapter 2.

14. See Jacob, note 7 above, chapter 5; and Carter, note 5 above, chapter 5.

15. See Jacob, note 7 above; and J. Flynn and P. Slovic, "Yucca Mountain—A Crisis for Policy: Prospects for America's High-Level Nuclear Waste Program," *Annual Review of Energy and the Environment* 20 (1995): 83.

16. Carter, note 5 above. Carter himself continues to support Yucca Mountain as a repository site, periodically bringing forward revised reasons for his position. See L.J. Carter, "Ending the Gridlock on Nuclear Waste Storage," *Issues in Science and Technology*, Fall 1993, 73; and L.J. Carter, "It's Time to Lay This Waste to Rest," *The Bulletin of the Atomic Scientists*, January/February 1997, 13.

17. National Research Council/National Academy of Sciences, Commission on Geosciences, Environment, and Resources, *Rethinking High-Level Radioactive Waste Disposal: A Position Statement of the Board on Radioactive Waste Management* (Washington, D.C.: National Academy Press, 1990).

18. Ibid., pp. 2–6.

19. Secretary of Energy Advisory Board, Task Force on Radioactive Waste Management, *Earning Public Trust and Confidence: Requisites for Managing Radioactive Wastes* (Washington, D.C., 1993). On DOE's management record, see Congressional Budget Office, *Cleaning Up the Department of Energy's Nuclear Weapons Complex* (Washington, D.C., 1994). An overview is provided in Flynn et al., note 7 above; and Flynn and Slovic, note 15 above.

20. U.S. Department of Energy, *Site Characterization Progress Report: Yucca Mountain, Nevada*, no. 12, DOE/RW-4077 (Washington, D.C., 1995), section 1.2.4, 1-6 to 1-8; "The Proposed Program Approach from the Office of Civilian Radioactive Waste Management," *OCRWM Bulletin*, DOE/RW-00448, Summer/Fall 1994, 1, 3; and Flynn and Slovic, note 15 above, pp. 104–06.

21. DOE is currently prohibited from developing an interim storage facility in the same state as a potential repository site, supposedly to help assure an objective assessment of a repository site. However, the prohibition can be changed by Congress, as was proposed in 1996.

22. R. van Konynenburg, "Gaseous Release of Carbon-14: Why the High-Level Waste Regulations Should Be Changed," in *High-Level Radioactive Waste Management: Proceedings of the Second Annual International Conference* (La Grange, Ill.: American Nuclear Society & American Society of Civil Engineering, 1991), 313–19.

23. H. Fabryka-Martin et al., *Summary Report of Chlorine-36 Studies: Systematic Sampling for Chlorine-36 in the Exploratory Studies Facility*, Level 4 Milestone Report 3783AD (Los Alamos, N.M.: Los Alamos National Laboratory, 1996).

24. K. Rogers, "Nuke Waste Exec Cautions Scientists on Chlorine Reports," *Las Vegas Review-Journal*, 30 April 1996, 2B; and Fabryka-Martin et al., note 23 above.

25. W.J. Broad, "A Mountain of Trouble," *New York Times Magazine*, 18 November 1990, 37.

26. K. Rogers, "Yucca Site Faults Deeper Than Thought," *Las Vegas Review-Journal*, 24 May 1994, 1 B, 4B; and R. Monastersky, "Faults Found at Nevada Nuclear Waste Site," *Science News* 145 (April 1994): 310.

27. U.S. Department of Energy, *Environmental Assessrnent: Yucca Mountain Site, Nevada Research and Development Areas, Nevada*, Volume 1, DOE/RW-(00)73 (Washington, D.C., 1986), Section 1.2.3.2.

28. Geomatrix Consultants, Inc. and TRW, *Probabilistic Volcanic Hazard Analysis for Yucca Mountain*, BA0000000-01717-2200-00082, Rev. 0 (1996), 2–14.

29. R.A. Kerr, "A New Way to Ask the Experts: Rating Radioactive Waste Risks," *Science* 274 (8 November 1996): 913.

30. Ibid.

31. J.L. Younker et al., *Report of Early Site Suitability Evaluation of the Potential Repository Site at Yucca Mountain, Nevada*, SAIC-91/8000 (Washington, D.C.: U.S. Department of Energy, 1992).

32. See K. S. Shrader-Frechette, *Burying Uncertainty: Risk and the Case against Geological Disposal of Nuclear Waste* (Berkeley, Calif.: University of California Press, 1993).

33. L.J. Carter, "It's Time to Lay This Waste to Rest," note 16 above, page 14.

34. U.S. Office of Management and Budget, *Statement of Administration Policy, S. 1271—Nuclear Waste Policy Act of 1996* (Washington, D.C., 1996). The Clinton administration cited these revised standards as one reason for the threatened veto.

35. Ibid., attachment with EPA views, p. 4.

36. J. Cantlon, *Nuclear Waste Management in the United States: The Board's Perspective* (Arlington, Va.: Nuclear Waste Technical Review Board, 1996), 7.

37. Flynn et al., note 7 above, chapter 6; and M.A. Greber, E.R. Frech, and J.A.R. Hillier, *The Disposal of Canada's Nuclear Fuel Waste: Public Involvement and Social Aspects*, AECL-10712 (Pinawa, Manitoba, Canada: Whiteshell Laboratories, 1994).

38. H. Kunreuther, K. Fitzgerald, and T.D. Aarts, "Siting Noxious Facilities: A Text of the Facility Siting Credo," *Risk Analysis* 13 (1993): 301–18.

12

Yucca Mountain Is the Best Solution to Nuclear Waste Storage

Luther J. Carter

Luther J. Carter, an early proponent of establishing a national nuclear waste center in Nevada, is the author of Nuclear Imperatives and Public Trust: Dealing with Radioactive Waste.

Yucca Mountain provides the best solution of any yet proposed on how to deal with the storage of radioactive nuclear wastes. Unfortunately, the issue of nuclear waste storage has become mired in petty politics and the overzealous machinations of various environmental and advocacy groups. By creating a safe, sound national nuclear waste repository, the United States would encourage the rest of the world to follow its example, providing further insurance against environmental contamination and weapons proliferation.

Another frustrating chapter in the long history of efforts to deal with spent reactor fuel was written before the 104th Congress adjourned in October [1996]. The Senate had passed a bill in July to force the pace of the nuclear waste program. But this measure, caught up in intense controversy, was approved by a majority that would be too small to override a promised presidential veto even if the measure could clear the House, which it could not.

The spent fuel problem will thus be carried over into the next Congress. There is no immediate crisis, but the continued failure of the United States to resolve the nuclear waste problem amounts to what is arguably a default on a moral obligation. As the nation that started the nuclear endeavor, the United States should be leading the world in sequestering dangerous and highly radioactive fission residues and byproducts.

It is doubly important that the United States set the right example for the rest of the world because the plutonium created by power reactors—already totaling hundreds of tons of weapons-usable material—would

Reprinted from "It's Time to Lay This Waste to Rest," by Luther J. Carter, *The Bulletin of the Atomic Scientists,* January/February 1997. Copyright © 1997 by the Educational Foundation for Nuclear Science, 6042 South Kimbark, Chicago, IL 60637, USA. A one-year subscription is $28. Reprinted with permission from *The Bulletin of the Atomic Scientists.*

also be sequestered. As long as the fuel is not reprocessed, the plutonium it contains is rendered relatively inaccessible for weapons use by the spent fuel's lethal radioactivity.

The nuclear power industry, faced with growing accumulations of spent fuel at reactor stations, wants an interim storage facility at the Nevada Test Site, because a permanent geologic repository will not be licensed and built at Yucca Mountain (which straddles the southwest edge of the test site) before 2010 at the earliest.

The utilities argue that they are owed relief given that the Energy Department has already collected some $12 billion from their ratepayers for the Nuclear Waste Fund. They point to a recent U.S. Circuit Court of Appeals ruling that the Energy Department is indeed obligated by the Nuclear Waste Policy Act of 1982 to begin accepting spent fuel in 1998.

But in seeking to help the utilities, the Senate Energy and Natural Resources Committee has overreached. The bill it reported went too far in allowing exemptions and exceptions from environmental law and regulation. Moreover, the exposure standard proposed for the hypothetical "average individual" in the vicinity of Yucca Mountain was set at 100 millirems a year—in stunning contrast to the 25-millirem light-water reactor "fuel cycle" standard that covers operation of all 100 reactors and all chemical conversion and processing plants (only uranium mining and the transporting of uranium and spent fuel are outside this standard).

The state of Nevada, for its part, has tried to stir up opposition in the cities and towns along the routes that prospective spent fuel shipments would follow. Nevada Sen. Richard H. Bryan even speaks of a "mobile Chernobyl," which is about as blatantly false and cynical a representation as one hears around Washington these days.

The continued failure of the United States to resolve the nuclear waste problem amounts to what is arguably a default on a moral obligation.

The White House contends that siting an interim facility at the Nevada Test Site before Yucca Mountain has been accepted or rejected for a geologic repository would prejudice the repository siting decision. The White House cited a recent report by the Nuclear Waste Technical Review Board that makes a similar criticism. But the review board, in the main a useful advisory body drawn largely from academe, had failed—as the White House and perhaps most members of Congress itself have failed— to face the fact that the nuclear waste program is now making its last stand in Nevada.

There are two stubborn, irreducible realities at play, one political, the other technical.

Politically, the reality is that a rejection of Yucca Mountain on technical grounds would not lead to a reopening of the search for geologic disposal sites, a pained, fitful, two-decade endeavor that produced mainly protests and histrionics.

Congress, in its 1987 amendments to the nuclear waste act, narrowed the search for the nation's first repository to Yucca Mountain. It did so a year after the Reagan White House—in something of a political panic—summarily abandoned the controversial search in the eastern half of the country for sites for a second repository. It also brought to an end some equally controversial plans for siting an interim surface storage facility.

Today the nation simply does not have the political stomach for another search for sites, for either a geologic repository or a surface storage facility.

Technically, the reality is that even if other geologic sites were politically available, probably nothing could be gained by choosing one of them over Yucca Mountain. Experience with geologic site exploration in the United States and abroad shows that all sites present problems, complications, and uncertainties.

No site can offer permanent and absolute containment of all radioactivity. The one irrefutable advantage of geologic disposal over surface storage is that geologic repositories are less subject to the vagaries of natural forces and human intrusion, especially over long reaches of time after institutional control has been lost.

The Energy Department now appears genuinely optimistic about the ongoing investigation of Yucca Mountain, and the department expects to make a positive "viability" determination in 1998 as a step toward a formal license application in 2001. A much-needed political lift for the project may come as early as next spring, when an impressive 25-foot-diameter tunnel, now being bored through the mountain in a five-mile exploratory loop, is completed.

But repository performance studies are complex, and even the site's advantages can also present disadvantages. For instance, the fact that the proposed repository would be built in relatively dry rock, high above the water table, is in most respects all to the good. But it also means that oxygen is present and that a corrosive "oxidizing environment" may shorten the life of waste packages.

Today the nation simply does not have the political stomach for another search for sites, for either a geologic repository or a surface storage facility.

Whatever the site's disadvantages, they might be compensated for over time through various engineering strategies—improving repository design or providing even more durable waste containers. (Performance studies indicate that the robust double-walled spent fuel casks now planned would, under most credible assumptions, last 10,000 years. These casks are to have a three-fourths-inch-thick nickel alloy inner wall and a four-inch-thick carbon steel outer wall).

The Nuclear Waste Technical Review Board has suggested, provocatively if somewhat offhandedly, a "more realistic" incremental approach to repository development and licensing. As the years go by, plans would be revised from one stage to the next, with "assured retrievability" maintained for all spent fuel placed in repository tunnels. As the board recognizes, such a plan would not work without a nearby interim storage facility to accommodate plan adjustments and to receive any retrieved fuel.

But the board believes it may be 2001 before enough is known about Yucca Mountain to decide whether it is suitable for a repository, and it recommends vaguely that "other potential sites for both disposal and centralized storage be considered."

Creating a national nuclear waste center might strengthen, at least by example, the international nuclear nonproliferation regime.

The final Senate bill was tied provisionally to the 1998 site "viability" determination, but its default position was that, one way or another, construction of an interim storage facility would begin by the year 2001 at the test site, unless another site was found and "approved by law." The bill thus recognized squarely and realistically that failure of geologic disposal plans would leave surface storage at the test site as the only option—apart from spent fuel remaining indefinitely at some 70 nuclear power stations.

If the new Congress and the White House are to meet the nation's obligations in the nuclear waste field, they must establish a national center with facilities for interim storage and final disposal of both spent fuel from nuclear power generation and high-level waste from the past production of nuclear weapons. The only place to do this is at the Nevada Test Site and at Yucca Mountain.

The surface storage facility would receive spent fuel pending the availability of the geologic repository and later allow loading of the repository to proceed at a pace recommended by project scientists. It would also turn Yucca Mountain politics on its head by giving the state of Nevada a compelling incentive to cooperate in achieving a successful repository project.

Increasingly over the next few decades reactors will be reaching the end of their original 40-year operating licenses. Also, certain reactors may be deemed to have become prematurely uneconomic or unsafe. Creating a national nuclear waste center in Nevada would allow utilities to shut down and decommission reactors for whatever reason without fear that they will be left with spent fuel. The need for a permanent resting place for high-level waste generated by the clean-up of the nuclear weapons production complex is also becoming more pressing. The Savannah River Plant is already producing canisters of vitrified waste.

On top of everything else, creating a national nuclear waste center might strengthen, at least by example, the international nuclear nonproliferation regime.

Every ton of spent fuel contains some 10 kilograms of plutonium (the Nagasaki bomb contained 6.2 kilograms). If recovered by chemical reprocessing, this reactor-grade plutonium, though not ideal for the purpose, can be made into weapons. Each year, the 430 commercial power reactors that are scattered over 30 nations are producing 6,000 to 7,000 tons of spent fuel containing roughly 60 to 70 tons of plutonium.

The world is already awash in separated, weapons-usable plutonium. There are hundreds of tons of it—in deployed weapons, in weapons marked for dismantling, in scrap at the nuclear weapons production complexes, and in stockpiles at commercial fuel reprocessing plants, especially in France and Britain.

Ending plutonium separation and storing all spent fuel would be the sensible thing to do if the global nuclear enterprise could agree on it (which it cannot). But creating an international system for spent fuel storage (or perhaps a few multinational regional systems) to which countries could choose to commit their fuel for the indefinite future would be a step in the right direction. An International Atomic Energy Agency (IAEA) committee is now looking at the possibility of such a system.

Establishing a U.S. center for spent fuel storage at the Nevada Test Site, with IAEA inspectors present to insure accountability, might in itself encourage the Europeans and the countries of East Asia to create regional centers of their own.

Some spent fuel from abroad might go to the Nevada center from the beginning, including American-origin highly enriched uranium fuel now being returned from foreign research reactors. Also, the center could host a hands-on international laboratory for studies looking to spent fuel storage and disposal in desert regions with conditions similar to those in Nevada.

Despite many suggestions to the contrary, the nuclear waste problem is not physically or technically intractable. Although the problem is difficult and unpleasant politically, it is solvable if too much opposition is not stirred up all at once, as happened in the political debacle of 1986. If by the end of the next Congress the problem still has not met a proper response, it will be a shameful commentary on our inability or unwillingness to treat a serious matter seriously.

Organizations to Contact

The editors have compiled the following list of organizations concerned with the issues debated in this book. The descriptions are derived from materials provided by the organizations themselves. All have publications or information available for interested readers. The list was compiled on the date of publication of the present volume; names, addresses, phone and fax numbers, and e-mail and Internet addresses may change. Be aware that many organizations take several weeks or longer to respond to inquiries, so allow as much time as possible.

Center for Environmental Education Research (CEER)
7049 East Tanque Verde Rd., Suite 368, Tucson, AZ 85717
(520) 722-3335
e-mail: ceer@cei.org • website: www.cei.org

The CEER's mission is to improve the quality of teaching about the environment in U.S. primary and secondary schools by ensuring that students receive unbiased, environmental information based on sound science and economics. The group opposes the "top-down" approach of stricter federal laws and regulations to solve environmental problems, and also it opposes what it sees as a pro-EPA bias in school teaching and materials. The center publishes *Update* newsletters and *On Point* policy briefs.

Citizens for a Better Environment (CBE)
407 S. Dearborn St., Suite 1775, Chicago, IL 60605
website: www.cbemw.org

Founded in 1971, CBE's attorneys, scientists, policy analysts, engineers, and community organizers have developed strategies for improving and protecting environmental quality. CBE lobbyists work for more comprehensive regulation of toxic wastes and the protection of sites under threat from pollution. CBE publishes a quarterly newsletter, *The Environmental Review*.

Citizens for the Environment (CFE)
470 L'Enfant Plaza SW, Washington, D.C. 20024
(202) 488-7255

Citizens for the Environment is a lobbying group and think tank that supports free-market solutions to the issues surrounding toxic waste and the environment. The CFE opposed the Clean Air Act of 1990 and helped defeat Proposition 128, a California ballot initiative that would have increased regulation and control of toxic waste. The CFE declares that free enterprise and modern business methods have had a positive impact on the environment. CFE scientists have published research purporting to show that acid rain and global warming are shams perpetrated by the environmental movement.

Clean Sites, Inc.
901 N. Washington St., Suite 604, Alexandria, VA 22314
(703) 739-1200
e-mail: cleansites@aol.com • website: www.cdclark.com/cleansites

Clean Sites is a private, nonprofit organization dedicated to helping government agencies, private companies, and communities evaluate cleanup technologies and deal with contaminated sites.

Clean Water Fund
4455 Connecticut Ave. NW, Suite A300-16, Washington, D.C. 20008-2328
(202) 895-0432
website: www.cleanwaterfund.org

A national nonprofit organization, the Clean Water Fund advocates measures to ensure safe drinking water and promotes conservation of land and water resources. The groups "Watershed to Watertap" program educates communities on how to protect their drinking water from wastes and toxins.

Coalition for Environmentally Responsible Economies (CERES)
711 Atlantic Ave., Boston, MA 02111
(617) 451-0927
website: www.ceres.org

A coalition of investor, environmental, religious, labor, and social justice groups, CERES advocates sound environmental business practices. Member and signatory businesses earn the group's endorsement by agreeing to the ten CERES principles, which provide guidelines for environmentally sound decisions and standardized public reporting of the environmental impact of their businesses. The group publishes the *CERES Reports*, annual environmental reports from companies that meet and endorse the CERES standards.

EnviroLink
5808 Forbes Ave., 2nd Floor, Pittsburgh, PA 15217
(412) 420-6400
e-mail: support @envirolink.org • website: www.envirolink.org

EnviroLink is an online resource that provides information, news, and educational materials and web links to discussion forums and online bookstores offering material on current environmental issues. *The EnviroLink News Service* is an online publication that provides EnviroLink users with the latest news and information about the global environmental movement.

Environmental Defense Fund (EDF)
257 Park Ave. S., New York, NY 10010
(800) 684-3322
e-mail: EDF@edf.org • website: www.edf.org

EDF is a nonprofit organization that promotes scientific research on pollution and offers bipartisan, efficient, and fair solutions to current environmental problems. The EDF publishes a monthly newsletter, available online and in hard copy, on current environmental policies and actions.

Environmental Protection Agency (EPA)
1200 Pennsylvania Ave. NW, Washington, D.C. 20460
(202) 260-2090
website: www.epa.gov

The Environmental Protection Agency is the federal agency that oversees enforcement of federal laws governing pollution and environmental hazards. The agency's stated goal is "to protect human health and to safeguard the natural environment—air, water, and land—upon which life depends." The EPA is mandated by law to draw up guidelines for the manufacture and use of toxic chemicals, and sets environmental regulations for manufacturers that use toxic and hazardous substances in the course of business. The EPA offers more than 5,000 publications, including the *Children's Environmental Health Yearbook.*

Friends of the Earth (FOE)
1025 Vermont Ave. NW, Washington, D.C. 20005
(877) 843-8687
e-mail: foe@foe.org • website: www.foe.org

FOE is a national, nonprofit advocacy organization dedicated to environmental and cultural issues. Favoring strict environmental protection and sustainable growth, Friends of the Earth organizes and catalyzes action on pressing environmental and social issues. The group also publishes the "Clean Water Report Card," "Atmosphere," "Paying for Pollution," and other pamphlets on the current status of various environmental issues.

INFORM
120 Wall St., New York, NY 10005
(212) 361-2400
website: www.informinc.org

An independent research and education organization that examines the effect of business practices on the environment and on human health. INFORM research into toxic waste was used as the basis of the Environmental Protection Agency's national database known as the Toxic Release Inventory. The group's reports and recommendations are used by government, industry, and environmental leaders to solve environmental problems. *INFORM Reports* is the group's quarterly newsletter.

Public Citizen
1600 20th St. NW, Washington, D.C. 20009
website: www.citizen.org

Founded by Ralph Nader, Public Citizen is a watchdog group and advocacy organization concerned with consumer safety, global trade issues, and environmental policy. The group publishes books, conducts research, initiates lawsuits, issues press releases, and holds public forums to support positions it holds as favorable to the individual consumer. Public Citizen publishes briefings on government policy and lawmaking, including several on nuclear safety and nuclear waste.

U.S. Council for Energy Awareness
1776 I St. NW, Suite 400, Washington, D.C. 20006
(202) 293-0770

The U.S. Council for Energy Awareness is a public relations agency established in 1980 after the nuclear accident at Three Mile Island, Pennsylvania. Supported by 400 power companies across the nation, the council promotes nuclear power as efficient and environmentally safe.

Bibliography

Books

Donald L. Barlett and James B. Steele — *Forevermore: Nuclear Waste in America.* New York: W.W. Norton, 1986.

Harold C. Barnett — *Toxic Debts and the Superfund Dilemma.* Chapel Hill: University of North Carolina Press, 1994.

Phil Brown with Edwin J. Mikkelsen and Jonathan Harr — *No Safe Place: Toxic Waste, Leukemia, and Community Action.* Berkeley: University of California Press, 1997.

Lee Ben Clarke — *Acceptable Risk? Making Decisions in a Toxic Environment.* Berkeley: University of California Press, 1989.

Mark Crawford — *Toxic Waste Sites: An Encyclopedia of Endangered America.* Santa Barbara, CA: ABC-CLIO, 1997.

Michael Gerrard — *Whose Backyard, Whose Risk: Fear and Fairness in Toxic and Nuclear Waste Siting.* Cambridge, MA: MIT Press, 1994.

Christopher Hilz — *The International Toxic Waste Trade.* New York: Van Nostrand Reinhold, 1992.

Richard Hofrichter, ed. — *Toxic Struggles: The Theory and Practice of Environmental Justice.* Philadelphia: New Society, 1993.

Valerie Kuletz — *The Tainted Desert: Environmental and Social Ruin in the American West.* New York: Dimensions, 1998.

Allan Mazur — *A Hazardous Inquiry: The Rashomon Effect at Love Canal.* Cambridge, MA: Harvard University Press, 1998.

Dixie Lee Ray — *Environmental Overkill: Whatever Happened to Common Sense?* Washington, DC: Regnery Gateway, 1993.

Fred Setterberg and Lonny Shavelson — *Toxic Nation: The Fight to Save Our Communities from Chemical Contamination.* New York: John Wiley, 1993.

Seth Shulman — *The Threat at Home: Confronting the Toxic Legacy of the United States Military.* Boston: Beacon, 1992.

Andrew Szasz — *Ecopopulism: Toxic Waste and the Movement for Environmental Justice.* Minneapolis: University of Minnesota Press, 1994.

Periodicals

Ivan Amato — "Can We Make Garbage Disappear?" *Time,* November 8, 1999.

John DeMont — "Sydney's Dangerous Legacy: Residents Confront a Toxic Nightmare," *Maclean's,* February 8, 1999.

David James Duncan   "The War for Norman's River," *Sierra*, May 15, 1998.

Joshua Karliner   "Earth Predators," *Dollars and Sense*, July 17, 1998.

Dan Kennedy   "Woburn Postcard: Civil Inaction," *New Republic*, March 15, 1999.

Richard A. Kerr   "Radioactive Waste Disposal: For Radioactive Waste from Weapons, a Home at Last," *Science*, March 12, 1998.

Tim Larimer   "Asia: Too Hot to Handle," *Time International*, October 11, 1999.

Loren McArthur and Marc Breslow   "Polluters and Politics," *Dollars and Sense*, July 17, 1998.

Jim Motavelli   "Toxic Targets: Polluters That Dump on Communities of Color Are Finally Being Brought to Justice," *E Magazine*, July 17, 1998.

Pat Phibbs   "Who's in Charge of Nuclear Waste?" *The World & I*, May 1, 1998.

Linda Rothstein   "Explosive Secrets," *Bulletin of the Atomic Scientists*, March 1, 1999.

Lynn Scarlett,   "The Promise and Peril of Environmental Justice," *Reason*, February 1, 1999.

Richard Stone   "Russia: Nuclear Strongholds in Peril," *Science*, January 8, 1999.

Laura Tangley   "World's Toughest Bugs," *U.S. News & World Report*, October 19, 1998.

Karen B. Wiley and Steven L. Rhodes   "From Weapons to Wildlife: The Transformation of the Rocky Mountain Arsenal," *Environment*, June 1, 1998.

Jim Wilson   "Putting Nuclear Waste to Work," *Popular Mechanics*, June 1, 1998.

Index